CONSERVATION OF SHARED ENVIRONMENTS

Environmental Science, Law, and Policy

Series editors

Marc Miller • Jonathan Overpeck • Barbara Morehouse

CONSERVATION
of Shared Environments

Learning from the United States and Mexico

edited by Laura López-Hoffman, Emily D. McGovern,
Robert G. Varady, and Karl W. Flessa

with forewords by Mark Schaefer and Exequiel Ezcurra

The University of Arizona Press   Tucson

The University of Arizona Press
© 2009 The Arizona Board of Regents
All rights reserved

www.uapress.arizona.edu

Library of Congress Cataloging-in-Publication Data
Conservation of shared environments : learning from the United States and
Mexico / edited by Laura López-Hoffman . . . [et al.] ; with forewords by Mark
Schaefer and Exequiel Ezcurra.
    p. cm. — (The edge, environmental science, law, and policy)
    Includes bibliographical references and index.
    ISBN 978-0-8165-2877-6 (cloth : alk. paper) —
    ISBN 978-0-8165-2878-3 (pbk. : alk. paper)
    1. Conservation of natural resources—Mexican-American Border Region—
International cooperation. 2. Environmental protection— Mexican-American
Border Region—International cooperation. 3. Mexican- American Border
Region—History. 4. United States—Foreign relations—Mexico. 5. Mexico—
Foreign relations—United States. I. López-Hoffman, Laura, 1973–
    S934.M59C66 2009
    333.72097—dc22

                                  2009033170

♻

Manufactured in the United States of America on acid-free, archival-quality
paper containing a minimum of 30% postconsumer waste and processed
chlorine free.

14   13   12   11   10   09   6   5   4   3   2   1

Edge Partners

Institute of the Environment
http://www.environment.arizona.edu
University of Arizona James E. Rogers College of Law
Hennegin Research Fund
Biosphere 2 and the B2 Institute
The Udall Center for Studies in Public Policy
The University of Arizona Press

Support for *Conservation of Shared Environments*

National Science Foundation (Award #0443481—Research Coordination
  Network: Colorado River Delta) for their support of workshops
National Science Foundation (Award #0409867) for their support of Laura
  López-Hoffman
University of Arizona College of Life Sciences and Agriculture
University of Arizona School of Natural Resources and the Environment

A Tribute

We pay tribute to Carlos Marin of the U.S. International Boundary Water
Commission and Arturo Herrera Solís of the Comisión Internacional de
Límites y Aguas, the commissioners for the United States and Mexico,
respectively, who tragically lost their lives in an airplane accident on September 15, 2008, while on a flood-reconnaissance mission to the Presidio-
Ojinaga border. The tragedy occurred as this book was being written. Both
Carlos and Arturo were the personification of the role and profile of a
commissioner: technically capable individuals, consummate diplomats who
were able to engage each other and their respective bureaucracies, consensus
builders, and true border residents who conveyed a bilateral perspective to
their national capitals. The need to forge sustainable agreements in binational water management—newer in vision, more holistic, and able to meet
the demands of the changing times—was clear to both men, especially in
the latter moments of their administration as they led binational discussions
on the future of the Colorado River and its delta. All of us who, like Carlos
and Arturo, care about U.S.–Mexico transboundary conservation feel a deep
loss. May they rest in peace.

Contents

Section 5   *Building Institutional Bridges*

Foreword: Transcending Political Boundaries

In 1890, John Wesley Powell, famed scientist and explorer of the southwestern United States, proposed that the West be organized around watersheds. The artificial boundaries of states, he reasoned, were counterproductive in dealing with the defining resource of the West: water. State boundaries fostered disputes and competition for resources. Instead, watersheds, as natural boundaries, would provide a framework for the logical development of policies, cooperation, and resource conservation. Although politically impractical, Powell's ideas were consistent with and perhaps laid the groundwork for the concept of integrated ecosystem management that would be articulated decades later.

Perhaps the greatest challenge of resource conservation is to find ways to preserve natural geographies and the complex biological and physical interactions that take place within them, given the constraining realities of artificial political boundaries and socioeconomic circumstances created by humans. Nature's boundaries evolved over billions of years. By comparison, human boundaries have emerged in a mere instant in time. Yet in that instant we have dramatically altered the landscape and the remarkable biodiversity that has adapted to it. The impact of state boundaries pales in comparison to the impact of the boundary between the United States and Mexico, particularly given today's socioeconomic realities.

In the late 1800s, when Powell was articulating the merits of organizing the West in a way consistent with nature's boundaries, what is now Arizona was a territory, heavily influenced by Mexico and a diversity of Native American tribes. In fact, prior to the Gadsden Purchase of 1853, what is today southern Arizona and New Mexico was northern Mexico. Over the past 150 years, the remarkable natural ecology of the environment shared by the United States and Mexico has been profoundly altered by the complex political ecology of human interactions within and across two nations. These interactions are often not limited to the immediate vicinity of the political boundary. Through the sharing of watersheds and migratory species, the common environment of Mexico and the United States extends far beyond the border, as may the causes and consequences of transboundary conservation challenges.

Conservation of Shared Environments is a compelling and articulate reminder, however, that even in such challenging circumstances, transboundary environments can be conserved if ingenuity, foresight, and creative collaboration are applied to overcoming the constraints and realities of political processes. Today, policymakers in both the United States and Mexico have a unique opportunity to work cooperatively with leaders in nongovernmental organizations, corporations, and academia to make a lasting contribution for the common good.

The collective insights and experiences of the authors of this outstanding work demonstrate that an abundance of such ingenuity, foresight, and even inspiration can be drawn upon to conserve and restore the unique landscapes that grace the border region of Mexico and the United States, to protect species that migrate beyond the border zone, and to sustain the ecosystem services that support the well-being of people in both countries. Whether serious initiatives emerge to conserve the extraordinary ecological connections between the United States and Mexico is not a question of scientific or technical capacity, but of the will of citizens and the government officials representing them to save something intrinsic to the culture and economy of environments that are on the verge of disappearing forever.

Mark Schaefer, Ph.D.
Director, U.S. Institute for Environmental Conflict Resolution
Deputy Executive Director, Morris K. Udall Foundation
Tucson, Arizona
September 2009

Foreword: The Keepers of This Land

In the 1980s, you could walk along the Sonoyta River in Sonora into Organ Pipe Cactus National Monument and quench your thirst in the springs at Quitobaquito. Sometimes I would meet the park rangers at the springs, and we would sit together under the shade of cottonwoods, sharing trivial facts about the plants and animals and minerals of the desert, and speculating aimlessly on the arrival of the next rain.

At that time, people from the Tohono O'odham Nation crossed the line freely. It was their land, all of it, from Caborca, Sonora, to Sells, Arizona. The tribe was there before Mexico and the United States existed; they greeted Father Kino when the first Jesuit missionaries arrived.

There was a feeling of community along the border, a sense of place. The border, the line that separates the two countries, was also what connected the two sides, or so it seemed at the time. In the peculiar dialectics of *la línea*, the border line was the element that brought the communities of both sides together.

Even in the late 1990s, people could just wade the Río Bravo at Boquillas, Coahuila, to reach the visitor's center at Big Bend National Park and then return to camp in the beautiful wilderness of Maderas del Carmen.

Alas, things have changed. The border is now very clearly what separates the two nations; it is a divide between the two cultures and a lacerating separation of common environments.

These reflections are what this book is about. It is about the environment shared by both countries; it is about nature, wildlife, ecosystems, and protected natural areas along the border. It is also about ecological connections that drive processes from deep inside each country—such as the nectar-feeding bats that follow the flowering of the giant columnar cacti from Oaxaca to Arizona, the hummingbirds that track the flowers of the ocotillos and the agaves along the slopes of the Sierra Madre, and the river-driven dynamics of the Colorado River delta. But it is also significantly about the hopes, anxieties, and expectations of human beings who work every day to protect our shared environments.

The volume is a written testimony of the state of cross-border conservation today. Within a scientifically rigorous perspective, the book puts together the work of many authors studying the biodiversity, natural

.resources, and shared environment of the United States and Mexico in a comprehensive baseline synthesis. The contributors are as diverse as the subjects, ranging from academia to government agencies to civil society organizations from both Mexico and the United States. They take a broad geographic approach to transboundary conservation, looking at continent-wide ecological connections as well as into more intimate environmental linkages in the borderlands.

This benchmark study will have historic value in the future because, as the book shows us, conservation is not only about plants and animals. It is also about people, about the keepers of the land.

Exequiel Ezcurra, Ph.D.
Director, University of California Institute for Mexico and the
    United States
Former President of Mexico's Instituto Nacional de Ecología
September 2009

CONSERVATION OF SHARED ENVIRONMENTS

Introduction

In honor of Cinco de Mayo in early May 2009, the newly elected president of the United States, Barack Obama, addressed a crowd of celebrants at the White House. Flanked by the Mexican ambassador to the United States, Arturo Sarukhan, the president spoke of the nature of the binational relationship between the United States and Mexico. His words touched on the historical, cultural, linguistic, and even gastronomic connections between the two countries. Speaking of the economic and security challenges shared by the nations, he remarked, "Good neighbors work together when faced with common challenges." In addition to the trials of commerce and security discussed by President Obama and Ambassador Sarukhan, the countries face another common challenge, one that has the potential to alter fundamentally the well-being of people in both nations. The challenge is to protect and conserve the environment shared by the United States and Mexico in the face of environmental change—drought, land-use change, intensive water use, deforestation, urbanization, habitat fragmentation, and climate change.

With this book, we aim to provide citizens and leaders in Mexico and the United States with a blueprint to work together to conserve their shared species, ecosystems, and ecosystem services. The chapters collected here—by authors from both countries and with backgrounds in academia, government agencies, and nongovernmental organizations—present a broad view about the challenges of conserving biodiversity across an international border. The authors bring to the book many years of experience working on transboundary environmental processes. In these pages, they examine the implications of their work for transboundary policymaking. We believe the results will be worthwhile and instructive to decision makers and other stakeholders and will resonate as well tomorrow as they do today.

Although the contributors to this volume describe many different types of transboundary conservation, common ecological themes run through the chapters: maintaining corridors and habitat for migratory species such as birds, bats, and butterflies in the face of land-use change; retaining habitat connectivity for species such as lizards, black bears, and Sonoran pronghorn antelopes whose natural range distributions are bisected by the political border and increasing physical barriers; preserving adequate water flows to support ecosystems and biodiversity in binational watersheds and landscapes under chronic water shortage and droughts; preserving the ecological integrity of borderland spaces in the face of the border wall, unilateral security, undocumented migration, and narcotics trafficking; and sustaining ecological processes that support the ecosystem services shared by people in both countries under a changing climate.

Throughout their chapters, the authors underscore the increased difficulty of addressing changes in the shared environment under the division of management between the two countries. The contributors show how when administrative regimes are divided, information sharing is slowed; decision making on projects that should be coordinated binationally can be delayed or disjointed among jurisdictions, yielding incoherent policies; and joint responsibilities for protecting shared resources may sometimes be overlooked. Several authors also describe how the usual problems that occur in any transboundary conservation situation may be exacerbated by the economic and cultural differences between the United States and Mexico.

As the species, spaces, and ecosystem services examined in these chapters vary in their geographic scope, so do the terms used to describe their respective settings. In our title and introduction to the book, we describe our subject as the shared U.S.–Mexico *environment* because this word is expansive both in concept and in geography. It conceptually encompasses species, ecosystems, and ecosystem processes as well as people and institutions. It is also sufficiently far-reaching geographically to include the continentwide ecological and social connections between the countries we consider in this book.

By the term *conservation*, we mean preserving species, ecosystems, and ecosystem processes and sustaining their contributions to human well-being. Throughout the book, we have taken *transboundary conservation* to mean any conservation effort that must consider ecological processes in another country as well as the cross-border consequences of actions, policies, and management regimes in one country for ecological processes in the other country. Under this definition, "transboundary" is not always "at or near the border." Because each chapter has its own central conservation problem, authors have at times used different terms to fit the scope of the problems they describe. In addition to *transboundary*, these terms include *transborder* or *cross-border* (usually referring to processes very close to the border) and *binational* (denoting the involvement of two national governments or peoples).

We have taken this broad geographic approach to transboundary conservation for two key reasons. First, species survival depends on connectivity of habitats in patterns that predate the contemporary political border. Many species have distributions that extend beyond the border zone.[1] Second, just as the ecological connections between countries may extend far across the political line, so may the causes, consequences, and solutions to transboundary conservation problems.[2] On-the-ground conservation activities and policymaking must therefore be focused at times far away from the border—for example, at critical bird stopover points throughout the Americas or within the vast Colorado River system far to the north of the river's delta in Mexico or on the ecological processes in Mexico that support life-sustaining ecosystem services in the United States and vice versa. In this sense, both the geographic scope and the topical scope of this book are more expansive than those of other volumes that have focused solely on the immediate U.S.–Mexico border region[3] or on shared ecosystems, ecoregions, and biomes.[4]

The volume's chapters are presented within five thematic sections. The first three sections are organized around the ways in which the United States and Mexico share their environment: spaces, species, and ecosystem services. "Conserving Shared Spaces" focuses on landscape conservation in the U.S.–Mexico borderlands. The second section, about animals that migrate and range *across* the U.S.–Mexico border, is entitled "Conservation of Shared Species." In the chapters of the third section, "Shared Ecosystem Services," the authors offer the novel approach of framing transboundary conservation through such services—the conditions and processes by which natural ecosystems support human life.[5] The fourth section, "Border Security and Conservation," discusses recent changes in border security and the effects of those changes on all species living in the border area. And the fifth section, "Building Institutional Bridges," hones our capacity for binational collaboration across the administrative differences between countries.

Read individually, the book's chapters provide insights into specific aspects of transboundary conservation, such as preserving habitat for migratory animals, protecting borderland wetlands, and sustaining the ecosystem processes that support shared ecosystem services. Read together as a whole, they reveal principles for how to frame successful transboundary conservation efforts: (1) the scope of transboundary conservation efforts may often need to extend far beyond the political line; (2) transboundary collaborations happen at many scales, from the local to the federal level, and should include diverse stakeholders; (3) informal, collaborative, and nongovernmental processes are instrumental in promoting transboundary conservation; (4) formal institutions are also crucial to establishing binational responsibilities for protecting shared environments and facilitating transboundary conservation; and (5) the concept of ecosystem services can be used to frame transboundary conservation in terms of mutual interests between countries.

After the individual chapters, we summarize the aforementioned ideas in "Guiding Principles for Successful Transboundary Conservation." In offering these principles, we suggest that the ideas are enduring—that they will be as relevant to U.S.–Mexico transboundary conservation efforts in the future as they are today. In "Energizing Transboundary Conservation—Three Easy Steps," we provide three simple, feasible policy recommendations that can be rapidly and easily implemented today. These suggestions reflect the guiding principles for framing transboundary conservation that emerge from this book and are actionable in the short term because they follow on existing binational institutional arrangements.

The publication of this book takes place at a time when a wall is being built to separate our two countries. Discussions of undocumented migration, narcotrafficking, and arms smuggling appear to be driving a deep wedge between us. The contributors to this book remind us, however, that we share an environment. They provide insights and tools for bridging the differences between our countries and addressing the impeding challenges of climate change, land-use change, and lost habitat. At this moment, it is crucial to remember how our environment unites us—how the rivers and aquifers, plants and animals, ecosystems and ecosystem services we have in common transcend the differences between us—and how as neighbors we would be well served to work together on the environmental challenges we share.

Notes

1. M. E. Soule and J. Terborgh, *Continental Conservation: Scientific Foundations of Regional Reserve Networks* (Washington, D.C.: Island Press, 1999).

2. L. López-Hoffman, R. G. Varady, K. W. Flessa, and P. Balvanera, "Ecosystem Services across Borders: A Framework for Transboundary Conservation Policy," *Frontiers in Ecology and the Environment* (2009), doi:10.1890/070216.

3. D. M. Liverman, R. G. Varady, O. Chávez, and R. Sánchez, "Environmental issues along the United States–Mexico border: Drivers of change and responses of citizens and institutions," *Annual Review of Ecology and Systematics* 24 (1999): 607–43.

4. C. Chester, *Conservation across Borders: Biodiversity in an Interdependent World* (Washington, D.C.: Island Press, 2006).

5. G. C. Daily, *Nature's Services: Societal Dependence on Natural Ecosystems* (Washington, D.C.: Island Press, 1997).

SECTION 1

Conserving
Shared Spaces

Conserving Shared Spaces

We begin our volume by considering the intimate connections of species, habitats, and people in the immediate vicinity of an arbitrary political line. These connections include the rivers, largely arid landscapes, and human histories that literally straddle the contemporary international border. The chapters in this section address many of the biological, social, and political complexities of managing the shared borderlands space between the United States and Mexico. They also demonstrate that drivers of change in borderlands environments may sometimes originate outside of the area.

In practice, the U.S.–Mexico borderland is not a fixed spatial concept. Paul Ganster and David Lorey explain how this region has been characterized in several ways, including as the 200-kilometer-wide (124-mile) swath defined in the 1983 La Paz Agreement between Mexico and the United States and as a border zone delimited by transboundary watersheds.[1] A culture-based definition of the borderlands might include a much broader space. Biological definitions, as demonstrated by the chapters in this section and elsewhere in the volume, can vary from a relatively narrow space inhabited by an indigenous plant species to a "borderland" that is nearly continentwide in the case of migratory birds. However, the chapters of this section contemplate transboundary conservation and policy specifically affecting spaces such as the cross-border Sonoran and Chihuahuan deserts, the riparian habitats of the Rio Grande/Río Bravo and the lower Colorado River, and contemporary Native American lands in close proximity to the political line.

The borderlands are at the historical root of many contemporary legal and social mechanisms for working on shared environmental problems between the two countries. For instance, the International Boundary and Water Commission/Comisión Internacional de Límites y Aguas was formed specifically to address the infrastructure and waters of the physical border. This history clearly emphasizes the importance of water allocation and water rights within the arid borderlands (an emphasis reflected in several of the chapters gathered in this section). Many of the first binational collaborations for conservation were established in this borderland zone; the chapters reflect the importance of continuing such collaborations.

The first chapter in this section, by Robert Varady and Evan Ward, traces an environmental history of the borderlands from the precolonial era to the present. It delves into the driving forces that have wrought environmental change in the borderlands, emphasizing the region's water-short character, changing water-management regimes, and binational institutions' role in the region over time.

The next two chapters center on contemporary questions of water management and how the region's two major hydrological systems sustain landscapes. In chapter 2, Francisco Zamora-Arroyo and Karl Flessa explain how overallocation of water, primarily within the United States, has devastated the Colorado River delta. They point out how a changing climate and increased water demand by cities will further reduce water supplies for the delta in the future. They suggest binational, cooperative strategies for restoring the delta. In chapter 3, Mark Briggs and his coauthors discuss the effects of water impoundment and invasive species on the riparian areas of the Big Bend reach of the Rio Grande. They provide a model for conducting small-scale, cross-border habitat restoration and long-term strategic planning to maintain the river system. Together, these chapters examine how efforts to restore wetlands and riparian areas in the shared U.S.–Mexico border region are impacted by patterns of human water consumption far removed from the border.

Next, in chapter 4, Rachel Starks and Adrian Quijada-Mascareñas describe how the legal status of Indigenous peoples in the United States and Mexico borderlands impacts their ability to participate in transboundary environmental decision making. They review a series of effective conservation partnerships between border tribes and other organizations on water, native plant resources, and border migration impacts. The authors recommend specific changes to multistakeholder, transboundary decision-making processes to incorporate tribal interests in conservation.

Like the section's first chapter, its final chapter, by Richard Knight, takes a historical view of borderlands spaces and environmental management. It entails a lyrical reexploration of Aldo Leopold's journeys through the Sierra Madre Occidental area in Mexico in the 1930s, highlighting Leopold's questions about the Apaches' influence on the area's natural history. The chapter further spotlights the close cross-border connections that have defined the shared spaces of the U.S.–Mexico borderlands over the past one hundred years.

Note

1. P. Ganster and D. E. Lorey, *The U.S.–Mexican Border into the Twenty-First Century*, 2d ed. (Lanham, Md.: Rowman and Littlefield, 2008), 12–13.

Chapter 1

Transboundary Conservation in the Borderlands

What Drives Environmental Change?

Robert G. Varady and Evan R. Ward

In a Nutshell

- Understanding dynamic environmental processes and their causes requires a solid appreciation of the spatial and temporal context within which these processes occur.
- Efforts to conserve a particular transborder environment are shaped by the natural and societal forces at play in that region.
- Social, cultural, legal, and administrative discontinuities on the two sides of the U.S.–Mexico border complicate joint efforts to protect a shared environment.
- The driving forces that most significantly affect conservation are ones that result in landscape changes.
- In a chronically water-short environment, historical land-use and water-use changes caused by human settlement, colonialism, economic development, public policies, and management regimes have had the most impact on habitat, species, and conservation.
- Natural phenomena such as drought, flooding, and climate variability and change—often exacerbated by human activity—also complicate transboundary conservation.

Introduction

The presence of political boundaries within continuous landscapes inevitably complicates all transactions that cross such boundaries. This chapter examines transboundary conservation in the U.S.–Mexico border area writ large—that is, the Sonoran and Chihuahuan deserts as well as the regions to the south and east of those deserts. Efforts to conserve a particular transborder environment are shaped in ways that depend on the specific natural and social forces at play in that region. So, for example, although certain commonalities may exist between regions, transboundary conservation in the dry, water-short, thinly populated U.S.–Mexico area faces different challenges than it would in the humid, densely populated Mekong Delta region. In short, understanding how borders affect biological conservation requires an appreciation of context.

As everywhere, climatic and physiographic processes on a geologic timescale have determined the overall features of the border region. Over the past few centuries and especially in recent decades, however, historical, human-induced processes—intensifying over time and acting in unison—have most strongly altered the environment.[1] Anthropogenic forces have also constrained societal attempts to conserve the environment and its resources. Aldo Leopold, a pioneering ecologist and enduring idealist, betrayed a deep distrust of these processes, observing that "many of the . . . forces inside the modern body-politic are pathogenic in respect to harmony with land."[2] The border region's present-day ecosystems offer a fitting backdrop for Leopold's assessment. Mexican, U.S., and Native American responses to environmental issues and conservation have been influenced by a host of interrelated driving forces. In this chapter, we discuss the drivers we consider most prominent: cultural heritage, demographics and economic development, and institutions.

Cultural Heritage of the Border Region

Indigenous Legacies Affecting Environment and Land Use

The border region and its hinterlands have been occupied by humans for at least ten thousand years,[3] and in the intervening millennia the area has accommodated numerous cultures. Indigenous peoples were the first to inhabit the area, beginning with prehistoric Paleoindian groups such as the Clovis culture and including the Uto-Aztecan peoples (e.g., Tohono O'odham, Yaqui, Tarahumara), settled nomadic (mostly Athabascan) tribes, Plains Indians (e.g., Comanche), Algonquians (e.g., Kickapoo), and other ethnic groups (e.g., Seri) and their descendants. Today, there are more than thirty American Indian tribes on the U.S. side of the border (twenty-one of them officially recognized), and about fifteen—although lacking official status—on the Mexican side (for more on these tribes' role in transboundary conservation, see chapter 4 in this volume).

These cultures have continuously affected the environment, manipulating the landscape and using its resources. There is archaeological evidence of prehistoric crop domestication and likely human-induced animal extinction. Canals uncovered in Arizona and New Mexico evince early attempts at irrigation. Game animals such as elk, antelope, deer, and javelina were hunted for food and skins; cactus fruits and other edible plants were collected and consumed; crops such as maize, which originated in what is now central Mexico, spread northward and southward and were commonly cultivated in the region more than three thousand years ago; rivers and coastal areas were fished; wood was cut, assembled, and transported for construction and fuel; and rocks, sand, clays, and ores were extracted for building, decoration, and household use. Indigenous peoples in the past relied almost exclusively on harnessing and obtaining resources from the land. There is evidence of these indigenous societies' careful stewardship of land, water, and wildlife, as well as of a worldview that included sustainability. For example, Apaches employed sophisticated ecological management systems to ensure that they left as light an imprint on the land as possible.[4] However, examples of extinction, denudation, overexploitation, and warfare are also plentiful. The impacts of these latter practices were likely limited by transportation and communication barriers, lack of large machinery, susceptibility to climatic extremes and natural disasters, and small and stable populations.

What is important for our purpose is to acknowledge that earlier indigenous peoples, like all human societies, used available resources to survive. Doing so often necessitated large-scale land transformation such as deforestation, conversion of grasslands to agriculture, stream diversion, and mining. These actions were in many cases accompanied by significant changes to species populations via habitat conversion or destruction; elimination of predators, "pests," and "weeds"; reduction of herd size; and overfishing. The present environment is a partial product of these actions.

Colonial Influences

The late-sixteenth-century advent of Europeans led by Spanish conquistadors and missionaries caused further land conversion.[5] New Spain—as the area was then known—became an important agricultural region. The new settlers, though few, brought transformative technologies, practices, and implements for construction, transport, agriculture, irrigation, and mining—all underlain by European concepts, management systems, and institutions.

These tools amplified humans' ability to alter the land, its resources, and its productivity.

During the Spanish colonial period (which lasted until Mexican independence in 1821), the Spaniards, employing local labor, raised livestock, introduced and grew wheat, established a land-tenure system, developed water resources, and began commodifying and "privatizing" resources—that is, providing private rights to access or outright ownership. The colonists raised cattle and other livestock and spread them widely. They also burned areas for additional grazing and cleared land for agriculture, garden plots, and towns; and like the indigenous inhabitants, they cut trees for construction and fuel, and rechanneled streams for irrigation. In addition, pursuing their perpetual search for El Dorado, the Spaniards dug mines and pits, finding some gold, but settling for silver, copper, lead, semiprecious stones, and rock and sand for construction. By the late eighteenth century, Sonora experienced mining booms that would recur over the next two centuries. These booms attracted residents from elsewhere in New Spain. The growing population impacted the natural-resource base through logging, charcoal production, and grazing.

Mining and grazing strengthened the economy, but they also brought overgrazing, deforestation, soil and gully erosion, aquifer depletion, stream contamination, habitat modification, wildlife loss, and urbanization. The impacts of colonial activities were further amplified by innovations in labor organization: religious salvation and education for Native peoples in exchange for labor on church-owned farms and workshops; centralized administration, featuring a strong mission-based paternalism; and a legal legacy rooted in the Iberian concept of community rights over individual rights. The Spanish legal system in the New World yielded policies that stipulated the distribution of scarce natural resources as well as of land and water rights in dry northern Mexico and what was later designated the southwestern United States.

In short, during three centuries of control and influence the conquistadors, missionaries, settlers, and merchants modified land, as the Indigenous peoples had, but in general with greater intensity and on a larger scale. Nonetheless, the region remained relatively thinly populated because of its climatic extremes, water scarcity, distance from centers of government, and political insecurity. This characteristic began to change by the mid–nineteenth century as the current boundary was being settled and further populated. The 1848 Treaty of Guadalupe Hidalgo and the 1853 Gadsden Purchase produced the present boundary line.

By then, the United States was experiencing "Manifest Destiny," and newly independent Mexico was seeking to exploit as yet untapped resources in its northern region. Government and private investors in both countries built railroads and developed water resources. These actions facilitated movement of people and goods and prompted settlements in once-remote areas. This growth was accompanied by expansion of irrigated agriculture, ranching, and mining—all employing more intensive and efficient modes of production. New mining boomtowns such as Tombstone and Bisbee in Arizona and Cananea in Sonora arose and quickly grew. At the same time, large tracts were cleared and set aside for grazing. The environmental legacy of mining and livestock, whose impacts were already palpable in the eighteenth century, included extensive deforestation around the mines and soil erosion in overgrazed areas.[6] The traditions and impacts of cattle raising and mining remain important today.

Border-Region Demographics and Economic Development

The most significant drivers of environmental change that have established conditions for present conservation were brought on by changing population and economic status. Water-ownership and allocation regimes in the two countries have been arguably the most influential economic drivers. Early in the development of the western U.S. frontier, surface and groundwater were allocated according to the "prior appropriation" doctrine. In Mexico, by contrast, water rights were federally held and provided to large commercial landholders as well as to communities, small holders, and *ejidos* (government-owned lands having communal usufruct rights). Already by the mid-1800s, during the transition from Spanish and Mexican legal systems to the U.S. one, these two modes of water ownership and allocation spawned widely different water-management regimes in those areas that had been part of Mexico but then became part of the United States. Historian Norris Hundley sees this transition as a strong impetus for a rise in unfettered individualism, leading to less attention to environment and common-pool resources on both sides of the border.[7]

By 1900, the population of the border states and territories had grown to about six million inhabitants (that number has more than doubled since then).[8] Thanks to continued federal investment in infrastructure, the early twentieth century was prosperous on the northern side of the border, but economic development was delayed in Mexico until after the Revolution of

1910–1920. Up to the 1960s, the agricultural sector experienced the greatest economic growth, especially through expanded irrigation—encouraged and supported by the two governments and enabled by large-scale land clearing in both countries. Mexico dammed major rivers such as the Conchos and Yaqui, established large irrigation districts, and developed groundwater resources in its north. Across the border, the United States also dammed tributaries to the Colorado River and the Rio Grande, permitting irrigated farming. Together, the two nations built two major dams—Falcón and Amistad—to manage the Rio Grande/Río Bravo.

During this time, many U.S. agriculturalists obtained their water rights and—in spite of high-evapotranspiration conditions—grew cotton, alfalfa, fruit, and vegetables in California and Texas. In Mexico in the 1950s, the "Green Revolution" introduced hybrid seeds and chemicals, which augmented yield and ushered in food-processing industries. Modern farming techniques that rely on heavy fertilizer and pesticide application, which was already widespread in the United States, introduced new risks to human and animal populations, streams, and habitats.

In the early twentieth century, two developments impacted land use and environment. In both nations, growing populations needed more energy, and the governments accordingly invested in hydroelectric power—the largest U.S. projects being the Boulder and Hoover dams on the Colorado River. To the east, dams were constructed on the Rio Grande, with Elephant Butte (1916) and Caballo (1938) the major ones to supply electricity. In Mexico, several large dams, most notably La Boquilla (1916), were built on the Río Conchos. Meanwhile, petroleum use grew with the advent of automobiles, and at the turn of the twentieth century the U.S. private sector accelerated its efforts to develop wells in Texas and California. Both dam construction and well-field development entailed substantial land-use changes, requiring earthworks, building materials, and substantial quantities of water. They also demanded new transportation networks, supply sites, and labor—all further altering ecosystems that were simultaneously impacted by population and development.

In Mexico, at the height of the dam-construction era and during the New Deal across the border, President Lázaro Cárdenas (1934–40) launched a major economic-reform project that redistributed land to peasant groups and created border free-trade zones. His government established import tariffs and incentives to expand manufacturing capacity. And in response to demand associated with World War II, industrial development grew sharply in northern Mexico, particularly in the state of Nuevo León. Military ports and bases concurrently became major employers. The presence of these facilities was often followed by defense manufacturing and services.

Population grew throughout the twentieth century, thanks to high fertility and movement of peoples—northward within Mexico and to the United States, and westward within the United States. As the area became industrialized and developed a service sector to support the new residents, cities such as Los Angeles and San Diego became metropolises. In the meantime, U.S. demand for cheap Mexican labor, especially in irrigated agriculture, drew migrants to the region. To fill wartime and postwar labor shortages in the United States, the Bracero guest farmworker program began in 1942, providing permits to more than four million agricultural Mexican workers.[9]

The successor Border Industrialization Program, which began in 1965, stimulated industrial development. It featured foreign-owned and foreign-based manufacturing and assembly plants (maquiladoras) located and operated in Mexico but permitted to export finished products back to the United States with reduced tariffs and trade barriers. As with agriculture, comparatively low costs—of labor, materials, and overhead—drew plants to the area, which produced textiles and simple manufactured goods at first but eventually automobiles, electronics, and chemicals. By 1990, more than twenty-five hundred such plants were operating, employing half a million workers and accounting for one-third of Mexico's trade.

Along with mines, whose wastes contaminated surrounding lands, airsheds, and watersheds, some of the maquiladoras were environmental polluters. Many of the production processes in electronics plants employed solvents and heavy metals, and they emitted hazardous waste and contaminated the air and water, particularly in the early years of their operation.[10] In addition, maquiladoras have had persistent secondary impacts by encroaching on habitat while converting open land to industrial facilities, heavily using water and other local resources for construction and operation, and drawing hundreds of thousands of migrants to communities without adequate safe water and sanitation.

All this economic activity—building infrastructure, supporting the war effort, expanding agriculture and industry, and supplying labor—has profoundly transformed the border environment in response to the needs of consumers throughout North America. Over the past half-century, these economic and demographic forces have radically altered land-use patterns and stressed the environment of this water-short region. Moreover, as Ramón

Ruiz has observed, the nature of the U.S.–Mexican economic relationship has been one of asymmetrical interdependence—characterized by transactions between a wealthy economy and a developing economy whose developmental hopes have been dependent on the growth of its northern neighbor.[11] The border region consequently became a consumer conduit, reaching deep into the United States with unforeseen ecological consequences.

Once the Bracero Program ended in 1964, legal immigration to the United States was severely curtailed, prompting large numbers of Mexicans to cross without proper documentation. Generally accelerating because the Mexican economy has remained weak, this process has caused serious friction between the two national governments. The immigration issue—already an irritant throughout the 1990s—became more acute after September 11, 2001, when the U.S. government conflated undocumented border crossings by economic refugees with crossings by individuals who have criminal or "terrorist" intent. A lengthy history of trade in contraband goods and services, abetted by the transnational border, further complicated the international border's commercial functions. The resulting problem was a commercial issue that was not included in the North American Free Trade Agreement (NAFTA): the trade in contraband products and its attendant criminal activity—particularly with regard to drugs and, to a much lesser degree, arms and exotic species, especially birds.[12] (See also the chapters in section 4 of this volume, regarding the impacts of border security on conservation efforts.)

Institutions of the Border Region and Beyond

The drivers discussed here have been important forces in reshaping the landscape within which transboundary conservation takes place. As we have seen, they also have continuously molded the societal environment by means of human movements and processes. But except in passing—particularly in noting the roles of water law and the Spanish missions—the foregoing material has not explored the importance of institutions in affecting the environment and setting the stage for conservation. Institutions are here interpreted broadly, including organizations, laws, treaties, compacts, sets of rules, established traditions, and other instruments—formal and informal, governmental and nongovernmental—aimed at organizing society.

Prior to the twentieth century, the most influential institutions were religious, governmental, and capitalist. Those that dealt explicitly with natural resources and water resources arose late in the nineteenth century, and nearly all were associated with governments—national, to a large degree, but occasionally regional and local (especially those that determined water-allocation regimes). The most prominent of these institutions was the International Boundary Commission (IBC), established by the Convention of 1889. Its responsibility was to define and maintain the integrity of the border between the United States and Mexico, but because half of the boundary was the Rio Grande/Río Bravo, the IBC was assigned authority over the rivers that defined and crossed the border. At first, concerned with demarcation and navigation issues, this commission had little to do with the environment. Because both countries drew waters from the rivers by constructing waterworks, however, the IBC's charge eventually grew to include matters of water quantity and quality, even while maintaining a strict focus on the line of demarcation. As the significance of binational water allocation increased, the IBC's role, scope, and responsibilities were redefined by the Water Treaty of 1944, which renamed it the International Boundary and Water Commission (IBWC). The 1944 treaty, which also defined Colorado River water rights for Mexico, assigned IBWC and its Mexican branch, the Comisión Internacional de Límites y Aguas (CILA), to adjudicate all treaty-related disputes. Over the decades, the IBWC/CILA interpreted its responsibilities strictly and was reluctant to move beyond its narrow border-only/water-only mandate.[13]

Until 1983, however, IBWC/CILA remained the only truly binational federal institution with any stake in the environment. Of course, within each country, there were moves to oversee, manage, and regulate aspects of the environment and natural resources. Legislation (e.g., the U.S. Mining Law of 1872) and the creation of federal U.S. departments such as Interior and Agriculture, their constituent agencies, and their Mexican analogues (e.g., for agriculture, forestry, and fisheries) created the infrastructure for such intervention. At a time when both governments sought to populate their frontier territories, however, many of these agencies saw their primary role as encouraging settlement, exploitation, and development; few saw land or water stewardship as priorities.

If domestic attention to environmental protection was rare, binational efforts were practically nonexistent until the adoption of the 1983 La Paz Agreement (also known as the Reagan–de la Madrid Accord),[14] which set up binational task forces on water and air pollution, natural resources and habitat, solid waste, and cross-border transportation of hazardous materials. Responding to calls for decentralization, the task forces included representatives of the environmental ministries and of the region's ten state governments as well as the diplomatic corps.

The La Paz Agreement has remained the bedrock of official U.S.–Mexico environmental cooperation. To implement its provisions, amendments have added several task forces and five new annexes that treat issues such as copper-smelter emissions and hazardous-waste shipment. The agreement also led directly to new binational protocols that established formal plans to improve environmental conditions. The Integrated Border Environment Plan (IBEP) was put in place in 1991 to address environmental objections to the trade treaty then under discussion; it was the first such program. It was criticized, however, for its lack of transparency, its omission of key areas of environmental concern, and its hazy, imprecise vision. Most serious was the plan's failure—in spite of its raison d'être—to acknowledge explicitly the environmental impacts of the then-proposed free-trade agreement.[15] Largely ineffective, IBEP was replaced during the Bill Clinton (1993–2000) and Ernesto Zedillo (1994–2000) administrations by the Border XXI Program. Heralded as the official heir to the La Paz Agreement, Border XXI joined the efforts of the U.S. Environmental Protection Agency (EPA) and the Department of the Interior with those of Mexico's environmental secretariat, then known as the Secretaría del Medio Ambiente, Recursos Naturales y Pesca (Secretariat of Environment, Natural Resources, and Fisheries). Border XXI convened working groups like those called for by the La Paz Agreement and commissioned databases on environment and natural resources. Given the politicized nature of these plans—IBEP and Border XXI were products of distinct administrations in each country—the George W. Bush (2001–2008) and Vicente Fox (2000–2006) governments unsurprisingly introduced a new plan, Border 2012. This plan stressed regional issues and reorganized the working groups along regional lines. After nearly two decades of post–La Paz plans, the chief criticism of this process has been that the plans have lacked the financial resources needed to succeed. Instead, each successive plan has relied on funds previously allocated for other purposes and has tallied the accomplishments of programs already in place.

Even as the La Paz process continued, another development arose. NAFTA spawned two environmental offspring—both prompted by pressure from environmental nongovernmental organizations (NGOs) wary of the negative impacts of increased trade. Simultaneous to NAFTA's adoption in 1993, an environmental side agreement was signed. This agreement established a trinational (Canada, United States, Mexico) commission to assure that each nation's environmental laws would be adhered to, even in the face of growing commerce. The Commission for Environmental Cooperation began functioning in 1994. Its effectiveness has been limited by small budgets and lack of power to enforce any findings (see also chapter 16 in this volume).

Also prompted by NAFTA, the presidents of Mexico and the United States signed a separate agreement to alleviate environmental problems in the border region by financing new infrastructure—primarily to improve water delivery and treatment.[16] The agreement established two sister institutions. One, the Border Environment Cooperation Commission (BECC, Comisión de Cooperación Ecológica Fronteriza in Spanish) was to solicit development projects from disadvantaged border communities and to guide them through a process leading to financing. BECC's design explicitly acknowledged the influence of community participation, calling for public input every step of the way. It also encouraged projects such as water-treatment plants to be environmentally and economically sustainable. The commission was to certify each project so that it could receive funding through the efforts of its partner institution, the North American Development Bank (NADBank), which was established by the same presidential accord. NADBank was charged with securing funds to lend, not give outright, to communities to build their BECC-certified projects. Over their first decade, BECC and NADBank certified and funded more than one hundred border environmental infrastructure projects at a total investment valued at U.S.$2.5 billion, while also providing a half-billion dollars from the U.S. EPA through its Border Environment Infrastructure Fund. In 2006, the BECC and NADBank boards of directors were merged.[17]

Numerous other institutions have addressed environmental problems and conservation, though few have been able to do so binationally. In the United States, the Good Neighbor Environmental Board (GNEB)—a group of persons representing federal and state agencies, private citizens, NGOs, and the private sector—has played a useful role in identifying and reporting important issues. Excellent examples are the GNEB's ninth report, on cultural and natural resources, and its tenth report, on environmental protection and border security.[18] In Mexico, a similar function is carried out by regional environmental *consejos*, appointed to advise the Mexican environmental secretariat (Secretaría de Medio Ambiente y Recursos Naturales) and comprising citizens and public servants.

In addition to formal governmental institutions, civil society groups and NGOs have been critically important in promoting cooperation for transboundaryconservationintheborderregionandbeyond(see,e.g.,chapter 2 in this volume). In many cases, these actors are involved in binational U.S. and Mexican conservation efforts, perform critical on-the-ground conservation work advocating new, responsive policies, and are the most highly motivated stakeholders to work in partnership across the border.

Conclusion

In this chapter, we have tried to demonstrate that landscapes and environments are ever-changing features that respond to physical, biological, climatic, and, perhaps most of all, human forces. The drivers of these changes are key determinants of how land appears, how it is used, and how it can be protected from degradation. In the case of the U.S.–Mexico borderlands, the key determinant of human and animal settlement, land use, economic development, and environmental transformation has always been the limited availability of water. The region's aridity to semiaridity has determined the nature and number of its flora and fauna and has strongly shaped social and institutional attempts to accommodate human settlement and economic development. The resulting place-specific patterns of land and water use have in turn influenced how governments, communities, and civil society have attempted to conserve landscapes, water flows, habitats, species diversity and populations, and environmental values.

Notes

1. D. Liverman, R. G. Varady, O. Chávez, and R. Sánchez, "Environmental issues along the U.S.–Mexico border: Drivers of change and responses of citizens and institutions," *Annual Review of Energy and the Environment* 24 (1999): 607–43. See also "Agreement Between the United States of America and the United Mexican States on Cooperation for the Protection and Improvement of the Environment in the Border Area" (La Paz Agreement), signed on August 14, 1983, full text available at http://www.epa.gov/usmexicoborder/ docs/LaPazAgreement.pdf/.

2. L. B. Leopold, ed., *Round River* (New York: Oxford University Press, 1993; orig. pub. 1972), 153.

3. B. J. Morehouse, D. Ferguson, G. Owen, A. Browning-Aiken, P. Wong-Gonzales, N. Pineda, and R. G. Varady, "Science and socio-ecological sustainability: Examples from the Arizona-Sonora border," *Environmental Science and Policy* 11(3) (2008): 272–84.

4. I. W. Record, *Big Sycamore Stands Alone: The Western Apaches, Aravaipa, and the Struggle for Place* (Norman: University of Oklahoma Press, 2008); personal communication with the authors, 2008.

5. This discussion draws substantially from the section "Environmental History" in Liverman et al., "Environmental Issues along the U.S.–Mexico Border."

6. G. L. Webster and C. J. Bahre, eds., *Changing Plant Life of La Frontera: Observations on Vegetation in the United States/Mexico Borderlands* (Albuquerque: University of New Mexico Press, 2001).

7. N. Hundley Jr., *The Great Thirst: Californians and Water, a History*, rev. ed. (Berkeley and Los Angeles: University of California Press, 2001).

8. P. Ganster and D. E. Lorey, *The U.S.–Mexican Border into the Twenty-first Century*, 2d ed. (Lanham, Md.: Rowman and Littlefield, 2008).

9. W. Cornelius, *Mexican Migration to the United Sates: The Limits of Government Intervention* (San Diego: Program in U.S.–Mexican Studies, University of California Press, 1981).

10. R. G. Varady, P. Romero Lankao, and K. Hankins, "Managing hazardous materials along the U.S.–Mexico border," *Environment* 43(10) (December 2001): 22–36; R. Sánchez, "Health and environmental risks of maquiladora in Mexicali," *Natural Resources Journal* 30(2) (1990): 163–86.

11. R. E. Ruiz, *On the Rim of Mexico: Encounters of the Rich and Poor* (Boulder, Colo.: Westview Press, 2000).

12. The U.S. Fish and Wildlife Service has noted that "even though there are no reliable statistics on the number of birds or other wildlife that illegally enter this country each year from Mexico, a 1997 study by the World Wildlife Fund/Traffic USA concluded that the 'southern border is . . . probably the most widely used route for illegal importation of parrots into the United States.'" See S. Cleva and P. Fischer, "Federal agents target illegal bird trade," U.S. Fish and Wildlife Service news release, 1998, available at http://www.r6.fws.gov/pressrel/98-20.htm, accessed July 8, 2009.

13. S. P. Mumme, "Engineering diplomacy: The evolving role of the International Boundary and Water Commission in U.S.–Mexico water management," *Journal of Borderlands Studies* 1(1) (1986): 73–108; International Boundary and Water Commission (IBWC), *History of the International Boundary and Water Commission* (El Paso, Tex.: IBWC, 2009), available at http://www.ibwc.state.gov/About_Us/history.html, accessed July 8, 2009.

14. See note 1 for a full citation to the La Paz Agreement.

15. R. G. Varady, "Are EPA and residents of the U.S.–Mexico border speaking the same language?" comments presented at the U.S. EPA–Mexico SEDUE hearings on the Integrated Border Environment Plan, Nogales, Ariz., September 26, 1994, pp. 1–3.

16. "Agreement Between the Government of the United States of America and the United Mexican States Concerning the Establishment of a Border Environment Cooperation Commission and a North American Development Bank," signed November 16, 1993.

17. R. G. Varady, "The North American Free Trade Agreement (NAFTA) and environment in the U.S.–Mexico border region," in *Encyclopedia of Environment*

and Society, edited by P. Robbins (Thousand Oaks, Calif.: Sage, 2007), available at http://sage-ereference.com/environment/Article_n779.html; R. G. Varady, D. Colnic, R. Merideth, and T. Sprouse, "The U.S.–Mexico border environment cooperation commission: Collected perspectives on the first two years," *Journal of Borderlands Studies* 11(2) (Fall 1996): 89–119.

    18. Good Neighbor Environmental Board (GNEB), *Air Quality and Transportation and Cultural and Natural Resources*, Ninth Report of the GNEB to the President and Congress of the United States (English and Spanish) (Washington, D.C.: GNEB, 2006); GNEB, *Environmental Protection and Border Security on the U.S.–Mexico Border*, Tenth Report of the GNEB to the President and Congress of the United States (English and Spanish) (Washington, D.C.: GNEB, 2007).

Chapter 2

Nature's Fair Share

Finding and Allocating Water for the Colorado River Delta

Francisco Zamora-Arroyo and Karl W. Flessa

In a Nutshell ————————————————————————

- The Colorado River delta once covered 2 million acres (800,000 hectares) supporting wildlife, plant life, and indigenous peoples.
- Since the construction of the Hoover Dam in 1935, the delta has shrunk to 10 percent of its former area, currently supported by incidental water flows that will likely disappear in the near future.
- The Colorado River Joint Cooperative Process—composed of U.S. and Mexican government agencies, nongovernmental organizations, water users, and water managers—is tasked with finding dedicated water sources for the delta while meeting urban and agricultural water needs.
- The options for acquiring a dedicated water source are undiverted Colorado River water, operational loss water (in Mexico), agricultural return flow, treated municipal effluent, naturally occurring groundwater, water from conservation actions, and acquisition of new water rights.
- The best options to ensure the delta's survival appear to be agricultural return water, municipal effluent, and acquisition of new water rights—each requiring collaboration and creativity by stakeholders to achieve.

Introduction

Along its 2,250-kilometer (1,400-mile) journey from the Rocky Mountains in the United States to the Gulf of California in Mexico, the Colorado River not only carved the Grand Canyon but also created one of the most important estuaries in the world, the Colorado River delta. Before the construction of dams, the delta supported nearly 800,000 hectares (2 million acres) of wetlands and abundant wildlife. The wetlands provided Indigenous tribes with materials for food, cultural ceremonies, shelter, and clothing. In the past hundred years, the delta has been reduced to less than 10 percent of its original size as more than 99 percent of the water that originally reached it has been diverted to support agriculture and cities in the United States and Mexico.

The "Law of the River"—the collection of interstate agreements, policies, and international treaties that governs the distribution of the Colorado—gives equal priority to agricultural and domestic water use and lesser priority to power generation and navigation. "Water for nature," or instream flows, is not recognized.

Today, the wetlands survive on "inefficiencies" in the water system—inadvertently released water that is not used by agriculture or cities—amounting to less than one percent of historical flows. Despite the reduction in water, remnant wetlands still support more than 371 species of birds,[1] and the Kwapa Tribe continues to maintain a close connection to the river.

The incidental water supplies that currently support the delta may no longer be available as Mexico and the United States more "efficiently" use water in response to current drought conditions. Both countries are implementing projects to reduce water losses in the agricultural and municipal sectors and during transportation. Unfortunately, such projects may lead to an "efficiency paradox" as the leaking canals and agricultural return flows that maintain the delta wetlands are eliminated.[2]

The current extended drought in the Colorado River basin is ironically creating opportunities for protecting the delta, however. As stakeholders in both countries develop creative strategies for adapting to drought, they are also discussing ways to secure dedicated water for supporting and restoring the delta. In the past two decades, nongovernmental organizations (NGOs), working collaboratively across borders, have broken down institutional barriers between countries and among water users to promote creative solutions for managing transboundary shared environments. As part of the recently formed Colorado River Joint Cooperative Process (CRJCP), NGOs are joining water users and managers from Mexico and the United States to develop binational projects for meeting municipal, agricultural, and environmental needs.[3] There is great optimism that this process will result in a new binational agreement ensuring protection of wetland areas in the delta.

This chapter describes the role of NGOs in the development of the CRJCP and then discusses options available to this binational group for securing water to sustain the delta's natural and seminatural habitats. We recommend possible sources of water for the delta and emphasize the binational responsibilities for preserving wetland habitats in the face of the region's continued rapid development and increasing demand for agricultural and municipal water.

The Delta's Resilience

With the construction of the Hoover Dam in 1935 and of Glen Canyon Dam in 1963, the United States built a reservoir capacity in Lakes Mead and Powell of four times the annual average flow of the Colorado River. The construction of Morelos Dam in Mexico in 1950 allowed for the diversion of Mexico's water allotment, leaving no designated flow for the Colorado River delta's ecosystems. During the seventeen years it took to fill Lake Powell, the river stopped flowing to the delta; by the 1970s, it was reduced from approximately 324,000 hectares (800,000 acres) of riparian corridors, wetlands, and estuarine areas to a few scattered patches of disconnected habitat.

In the 1980s and 1990s, with Lakes Mead and Powell nearly full, a series of wet years forced the U.S. Bureau of Reclamation to release water into the river. These releases flooded the delta's riparian corridor, bringing back riparian and estuarine areas. Newly vegetated areas of native trees created habitat for birds and other wildlife. Although dry conditions have returned during the 2000s, sporadic accidental releases and agricultural return flows from the United States and Mexico have maintained important wetland habitat.

Today the Colorado River delta is a critical stopover on the Pacific migratory bird flyway, with 371 bird species breeding, wintering, or migrating through the area, representing 55 percent of the total bird species in North America.[4] The delta is the winter home to more than 400,000 migratory waterbirds and permanent home to 70 percent of the total population of the Yuma Clapper Rail (*Rallus longirostris yumanensis*), a binationally protected marsh bird.[5] In addition, the floods of the 1980s and 1990s supplied significant freshwater to estuaries in the upper Gulf of California, enhancing important breeding and nursery grounds for diverse and economically important marine fisheries such as shrimp, finfish, and shellfish (see chapter 10 in this volume).

The remnant delta includes the Andrade Mesa wetland in Mexico, which is supported by seepage that flows across the border from the All-American Canal in California, and the Ciénega de Santa Clara, which receives agricultural drainage water from the Wellton-Mohawk Irrigation and Drainage District in Arizona (see fig. 2.1). The Ciénega de Santa Clara is home to the endangered Yuma Clapper Rail. The Andrade Mesa wetlands provide habitat for more than 100 species of birds, some of them listed as endangered or threatened.[6] The major threat to these wetlands is the lack of a dedicated source of water. The Andrade Mesa wetlands will receive significantly less water once the All-American Canal is lined (see chapter 9 this volume),[7] and the Ciénega de Santa Clara will be affected if a long-dormant desalting plant in Yuma, Arizona, is restarted.

The Formation of Binational Institutions for the Delta

In 2000, Minute 306 was added to the 1944 International Water Treaty between Mexico and the United States. The minute created a working group to coordinate scientific studies of the delta's ecosystems, but it did not make specific recommendations on securing water. Only government representatives were initially allowed to join the working group; NGOs joined in 2004. At first, the group avoided discussing potential water sources for the delta because of opposing views between stakeholders in both countries about who was responsible for the conservation of the delta. This avoidance precluded an open dialogue about ways to protect the delta.[8]

In the 1990s, Mexican and U.S. academic institutions and NGOs began collaborative studies in the delta with the goal of determining which habitats should be conserved and how much water would be needed. In 2005, the NGOs published *Conservation Priorities in the Colorado River Delta*, a report identifying important areas to conserve, possible water sources, and restoration options.[9] The Minute 306 fourth working group formally accepted the report in 2006 as a guide for future efforts to protect and restore the delta.

Around 2000, the water surpluses in the Colorado River basin ended, and U.S. basin states began to discuss options for dealing with water shortages. Stakeholders initially feared the water shortage would limit the opportunity to secure water for the delta, but that was fortunately not the case.[10] A group of NGOs expanded on the U.S. basin states' Intentionally Created Surplus (ICS) approach. The ICS program allows water contractors to store in Lake Mead any water that results from the reduction of their use of water (through water-conservation activities or other mechanisms). This conserved water can then be used in the future. The NGOs developed the

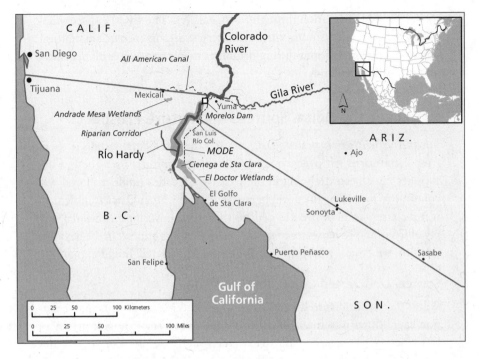

Figure 2.1. General map of the Colorado River delta in Mexico. The map shows some of the conservation priority areas (not drawn to scale). Map drawn by Mickey Reed.

notion of "taking ICS to Mexico," which would allow Mexico to prepare for water shortages by storing water in Lake Mead. Based on the success of their participation in the water-shortage discussions, NGOs subsequently initiated informal dialogue with water-management agencies in Arizona, Nevada, California, and Mexico. The enthusiasm associated with this informal dialogue generated a new *formal*, binational stakeholder process, the CRJCP, under the auspices of the International Boundary and Water Commission (IBWC) and its Mexican counterpart, the Comisión Internacional de Límites y Aguas (CILA). The CRJCP includes government agencies, water stakeholders, and NGOs from both countries (see chapter 17 in this volume for a discussion of IBWC/CILA).

After several months of developing the operational framework, the CRJCP began activities in March 2008 by identifying common goals and objectives; discussing and developing proposed collaborative projects to meet municipal, agricultural, and environmental needs; and recommending modes of implementation through IBWC/CILA.[11] The CRJCP successfully facilitated discussion among different water-use sectors (urban, agricultural, and environmental) about the opportunities to improve water management and meet their demands.

Water for the Delta: Sources and Future Prospects

Today, the delta's only water source is water that accidentally spills or seeps from canals or is left over from domestic or agricultural use. These water supplies have been declining in the past two decades (table 2.1) and will continue to decline as the drought continues and water agencies implement water-conservation projects to address shortages. Ensuring a dedicated supply of Colorado River water to the delta's wetlands is critical. In the next few sections, we describe options that the binational CRJCP might explore.

Source: Undiverted Colorado River Water

Water that is not diverted by Morelos Dam into Mexico's principal irrigation canal flows into the riparian corridor.[12] When the storage capacity of U.S. reservoirs is exceeded due to extremely high precipitation, some water flows to Mexico in excess of treaty allotment. This happened most recently in 1998–99. Excess water also crosses the border due to operational losses: water set to be delivered to U.S. farmers gets refused after upstream release because local rain made the water no longer necessary. Operational loss water reaches the delta if Mexico is unable to divert it at Morelos Dam. Undiverted water was the delta's largest single water source from 1998 to 2001 (table 2.1).

Prospect for the future

Excess flow from the United States is likely to be eliminated in the near future. Should reservoir levels recover after the current drought, the application of ICS will mean more water captured in upstream reservoirs, thus decreasing the probability of controlled releases downstream to Mexico. In addition, a small reservoir being built along the All-American Canal, "Drop-2," will prevent operational losses of water ordered by Imperial Valley farmers. According to the U.S. Bureau of Reclamation, this additional capacity "is needed to increase beneficial use of water released from Parker Dam in the United States to minimize unscheduled deliveries to Mexico."[13]

Allocations of Colorado River water, determined in the early 1900s, were based on an overestimate of the river's long-term flow. At the time, water

Table 2.1. Sources and volumes of water (in million cubic meters [mcm]) available for environmental purposes, 1998–2002.

Source	1998	1999	2000	2001	2002	Future
Undiverted flow	2,967.0	1,062.0	166.5	135.7	40.7	↓↓
Operational spills	297.3	132.0	87.6	30.8	12.3	↓↓
—Agricultural return flow—						
Ayala drain	22.0	16.9	22.0	16.9	19.0	↓
Drains to Río Hardy	16.0	13.9	13.9	8.0	4.9	↓
MODE Canal	139.4	97.0	132.8	127.9	129.5	↓?
Santa Clara drain	17.9	14.8	16.0	14.8	13.6	↓
—Effluent—						
San Luis	9.9	11.1	12.0	12.0	13.0	↓↓
Mexicali[a]	40.5	40.5	40.5	40.5	40.5	↑↑
—Groundwater—						
All-American Canal seepage[b]	83.5	83.5	83.5	83.5	83.5	↓↓
Total flow	3,593.5	1,471.7	574.8	470.1	357.0	↓↓
% virgin flow[c]	19.8	8.1	3.0	2.67	2.0	< 2

Note: Sources of flow estimates are from F. Zamora-Arroyo, J. Pitt, S. Cornelius, E. Glenn, O. Hinojosa-Hueta, M. Moreno, J. Garcia, P. Nagler, M. de la Garza, and I. Parra, *Conservation Priorities in the Colorado River Delta, Mexico and the United States* (Tucson, Ariz.: Sonoran Institute, Environmental Defense Fund, University of Arizona, Pronatura Noroeste Dirección de Conservación Sonora, Centro de Investigación en Alimentación y Desarrollo, and World Wildlife Fund–Gulf of California Program, 2005), available at http://sonoran.org/programs/sonoran_desert/si_sdep_delta_priorities.html#report. The following are exceptions:

[a] M. J. Cohen and C. Henges-Jeck, *Missing Water: The Uses and Flows of Water in the Colorado Delta Region* (Seattle: Pacific Institute, 2001).

[b] Imperial Irrigation District, *All American Canal Lining Project* (2006), available at http://www.iid.com/Water_Index.php?pid=64, last viewed December 4, 2006.

[c] Total flow as a percentage of the 18,124 mcm average annual flow estimated from tree ring variation from 1490 to 1997 (C. A. Woodhouse, S. T. Gray, and D. M. Meko, "Updated streamflow reconstructions for the upper Colorado River basin," *Water Resources Research* 42 [2006], W05415. We use their highest estimate, "Lees-C"; their lowest estimate, "Lees-D," is 17,656 mcm per year.)

Arrows indicate likely future trend: ↓ indicates likely decrease; ↓↓ indicates likely major decrease; ↑↑ indicates likely major increase.

managers did not realize that the Colorado River basin periodically experienced droughts.[14] As the river reverts to its long-term average conditions, decreasing water supplies will place an even higher premium on minimizing excess flows to Mexico. In addition, alterations in the hydrologic regime resulting from global warming will further decrease the likelihood of controlled releases.[15]

Source: Operational Losses within Mexico

Operational spills in Mexico are usually the result of overdeliveries from the United States. Mexico has no reservoir capacity to store operational losses; if the spills are small, they are captured in irrigation canals, Mexico's only reservoir capacity; otherwise, they are released back into the floodplain through waste ways (fig. 2.2). Mexican operational spills decreased during the 1999–2002 period (table 2.1) as a result of declining overdeliveries from the United States. Operational spills are critical for maintaining small riparian areas in the delta and locally recharging the aquifer.

Prospect for the future

Operational losses are likely to decrease due to increased storage capacity in the United States; increased efficiencies in Mexican river management, including several proposed small reservoirs; and increased demand for irrigation and municipal water in Mexico.

The CRJCP can apply the ICS approach to mitigating the loss of inadvertent water releases in Mexico. An ICS is a water credit that a stakeholder can create through a series of water-conservation or market mechanisms. Water is stored until the stakeholder requires it. The IBWC/CILA will have to approve the use of an ICS approach for water to be stored in U.S. reservoirs for Mexican stakeholders. Another option is the Colorado River Delta Water Trust, created by NGOs to secure regular, base-flow water through water purchases in the Mexicali Valley.[16] Approximately 1.7 million cubic meters (mcm) (1,376 acre-feet) of water have so far been acquired in the past two years for restoration of riparian areas; the NGOs hope to acquire an additional 63 mcm (51,000 acre-feet). The CRJCP may be able to support the water trust with additional investments and by facilitating the delivery of this water to the river and restoration areas.

NGOs would like to use ICS to secure pulse flows to mimic natural floods in the riparian corridor and estuary. ICS pulse flows would complement the base flow being secured through the water trust. Taking this step, however, would require a binational agreement allowing the transfer to

Figure 2.2. Waste way, or *desfogue,* located 27 km (17 miles) south of the U.S.–Mexico border at Morelos dam. The photograph shows water being released from the irrigation canal network back into the river on February 11, 2009, after a rain event in the Yuma–Mexicali Valley area. Photograph by Francisco Zamora-Arroyo.

Mexico of water purchased or secured through other mechanisms by NGOs in the United States and Mexico.

In order to put these ideas into action, it will be necessary to identify when and where water is most needed in the riparian areas and estuary. Mexico's extensive irrigation structure allows for delivery of water to practically any point along the river, but close coordination between water agencies and NGOs will be required to ensure that even small flows can be directed to the areas where it is needed to protect and restore the ecosystem.

Source: Agricultural Return Flow

Agricultural return flow is water that drains away from agricultural fields. Due to high salt content in water and soils, farmers often have to apply extra water to flush the fields. The agricultural return flow that results is thus typically higher in salinity (often as high as three parts per thousand) than undiverted river water or water from operational losses. Return flows show a decrease over the 1998–2002 period (table 2.1).

Prospect for the future

In the future, agricultural return flow will be reduced as water is moved from agriculture to urban uses. The cities of Mexicali and Tijuana have been acquiring agricultural water rights. Although they have enough water to meet short-term demand, they will likely continue buying agricultural water rights in the long term. An even greater reduction in return flow will result from increased agricultural efficiency. At present, agriculture in Mexico is not as water efficient as agriculture in the adjacent Imperial Valley of California. As economic incentives cause Mexican farming and irrigation practices to become more water-use efficient, environmentally beneficial return flows will decrease. In a recent study, the Comisión Nacional del Agua (National Water Commission) estimates that farms can increase water-use efficiency from 65 percent to 80 percent.[17] Although the report does not estimate a reduction of return flows, it is expected that they will decrease by 15 percent at least.

The brackish groundwater supplied to the Ciénega de Santa Clara in Mexico via the Main Outlet Drain Extension (MODE) canal from the Wellton-Mohawk Irrigation and Drainage District in Arizona is a special case. In the long term, the supply of this water depends on the economic viability of agriculture in the lower Gila River valley of Arizona. If irrigation water is diverted for municipal uses in Arizona, the MODE will carry less water because fewer fields will require pumping of saline groundwater. In the short term, MODE water may be used in the Yuma Desalting Plant, resulting in reduced flows and higher-salinity water reaching the Ciénega de Santa Clara,[18] and likely reducing wetland extent. Although the Yuma Desalting Plant's major stakeholders are trying to find ways to maintain the wildlife habitat and ecosystem values of the Ciénega de Santa Clara, water supply from the MODE is not assured. The case of a proposed test operation of the desalting plant and the Ciénega de Santa Clara is being discussed by the CRJCP and represents one of the first challenges for this binational process.

Of particular concern is the potential loss of return flows to the Río Hardy area, which represents the main source of water to the estuary. Securing existing return flows is currently very important as Mexican and U.S. NGOs attempt to restore estuary conditions in the area, but they need dedicated water sources. One promising option, in addition to existing return flows, is redirecting to the Río Hardy in Mexico 1.2 cubic meters per second (m³/s) (42 cubic feet per second [ft³/s]) of Mexican return flow that is currently going to the New River in the United States. It would cost U.S.$10 million to build the infrastructure needed to redirect the return flow from the New River to the Río Hardy.[19] This price is reasonable considering the significant ecological and economic benefits of restoring estuary conditions to the Hardy. This flow will double existing flows to the Hardy and increase the possibility that water will reach the Gulf of California. The high concentration of nutrients in agricultural return flows would increase the primary productivity of the restored estuary, benefiting fish and shellfish species.

Source: Municipal Effluent

The delta region includes two major cities, San Luis Río Colorado, Sonora (2000 census population 180,000), and Mexicali, Baja California (2000 census population 800,000), as well as many small farming communities where wastewater is either treated in septic systems or is released into surface waters. For years, untreated effluent from Mexicali was discharged into the New River, which flows north across the border and into California's Salton Sea. Today, nearly 1.0 m³/s (35 ft³/s) of the effluent is being treated in a new wastewater-treatment plant, Las Arenitas, significantly reducing the New River's pollution problem. In 2007, NGOs negotiated a twenty-year agreement with the state of Baja California to allocate 30 percent of the treated effluent (approximately 0.5 m³/s [17.5 ft³/s]) to the delta's Río Hardy for use in the river- and estuary-restoration project.[20] This is the first explicit allocation of water for nature in the Colorado River delta and represents a groundbreaking collaboration between NGOs and the state government of Baja California.

Prospect for the future

The supply of effluent water is likely to increase, although its availability for environmentally beneficial flows may decrease. Las Arenitas treatment plant in Mexicali is expected to double in capacity to 1.6 m³/s (56 ft³/s) by 2010. However, the state water authority is exploring treatment technologies to produce water for reuse in agriculture and industry. All 0.8 m³/s (28 ft³/s) of treated water from the plant is currently being sent to an artificial wetland for additional treatment before it discharges into the Río Hardy. However, future reuse of this effluent, possibly by agriculture or industry, might mean that Baja California will not be able to commit more water than the 30 percent (0.5 m³/s) already committed. The City of San Luis Río Colorado is now treating and reinjecting wastewater to the aquifer; untreated water was previously sent to the Colorado River. As San Luis and other delta towns and cities grow, more treated and untreated effluent water will be produced; however, it is uncertain whether this water will be dedicated to human reuse or to environmental flows to the delta.

Source: Groundwater

Natural springs are rare in the delta. Groundwater seepage from unlined irrigation canals supports some natural and seminatural habitats. The most notable example is seepage from the unlined portion of the All-American Canal that supports wetlands in the Andrade Mesa. By lining the canal, the U.S. Bureau of Reclamation will save 83.5 mcm (67,700 acre-feet) per year of water, but less water will support the wetlands.[21] Groundwater is pumped throughout the eastern half of the delta for irrigation. Michael Cohen and Christine Henges-Jeck estimate volumes at 777 to 950 mcm (630,000 to 770,000 acre-feet) per year.[22] Water derived from this source may support seminatural and natural habitats as agricultural return flow (discussed earlier) or as a contribution to river flows. An area being restored by NGOs is fed in part by seepage flows from adjacent farmland.

Prospect for the future

In the future, less groundwater is likely to be available. Lining of irrigation canals—both in the United States and in Mexico—will cause a reduction in the seepage that supports some wetland habitats. The most immediate concern is the All-American Canal; when the lining is completed, the water supply to the Andrade Mesa wetlands and groundwater for nearby Mexican agriculture will be reduced. The water table along riparian areas may fall as stakeholders overpump groundwater in response to shortages of surface water. Depths from the surface to groundwater may increase beyond the range suitable to native vegetation. Measures need to be taken to mitigate the risks of increased groundwater pumping. A detailed regional groundwater and surface model is required to estimate the trade-offs between water-conservation projects and aquifer recharge, on the one hand, and potential impacts to natural areas, on the other. This model will also help evaluate potential impacts from increased groundwater pumping for urban uses on riparian and natural springs areas.

Conclusions

The remaining wetland areas in the Colorado River delta depend on incidental water flows—mainly undiverted water and agricultural return flows. Future changes in river management in the United States and Mexico to increase water-use "efficiency" will most likely eliminate these incidental flows. The delta ecosystem will thus not survive if it continues to depend on incidental flows. Protection of the remnant 47,000 hectares (117,000 acres) of wetlands of the Colorado River delta in Mexico will require an allocation of water for environmental purposes and active management of that allocation to ensure that priority areas are sustained and high-value wetlands are restored. Water to sustain delta habitats in Mexico is a responsibility that must be shared with stakeholders in the United States.

The Colorado River Delta Water Trust has a goal of securing 63 mcm (51,000 acre-feet) of water for a base flow to the delta.[23] The trust has already acquired 1.7 mcm (1,376 acre-feet); it requires additional financial resources to reach its goal. The successful collaboration between NGOs and the state government of Baja California in securing treated effluent from Mexicali for environmental flows to the Río Hardy should be expanded. The commitment by the state government to dedicate 30 percent of effluent might be replicated by other agencies in Mexico and the United States.

The binational CRJCP must develop creative ways of securing both water and financial resources for on-the-ground restoration in the delta. State and federal governments in the United States and Mexico are fortunately now considering actively managing water for environmental purposes. In addition to securing water rights for nature, the CRJCP must also establish agreements to ensure that water can be reallocated and delivered in a timely fashion to priority conservation areas. For some of these mechanisms, such as ICS, it will be necessary to add a new minute to the 1944 Water Treaty to permit water stored in the United States to be used in Mexico both for a base flow to maintain water in the river continuously and for pulse or periodic flood flows.

The CRJCP's work provides an excellent foundation for the creation of a binational agreement. After a year of work, the CRJCP will begin to present suggestions for binational projects to the IBWC/CILA. Although project approval rests within IBWC/CILA, the participation of a wide range of stakeholders in the CRJCP will make it easier for governments to reach a binational agreement. It will take time and effort to establish the legal and administrative framework for managing water for environmental purposes in the Colorado River delta. The binational CRJCP discussion may be the best, if not the last, opportunity for both countries to define the delta's future.

Lessons drawn from the Colorado River delta and the CRJCP process can benefit people working on transboundary water policy in other regions. Although both federal governments formalized the CRJCP process (via the IBWC/CILA), it was originally the initiative of local, state, and regional stakeholders from both the United States and Mexico—stakeholders whose well-being will ultimately be impacted by water-management decisions and who are knowledgeable about needs and potential solutions. How these needs and interests will be reflected in the CRJCP's final products remains to be seen, but at least stakeholders have been part of the process. NGOs' role in ensuring the protection and restoration of the ecosystem has been paramount. Their participation has been very valuable not only in identifying areas to be protected and restored but also in proactively presenting creative solutions, from "taking ICS to Mexico" to the difficult challenge of equitably sharing transboundary water.

Acknowledgments

Karl Flessa is grateful for the support of the National Science Foundation (RCN 0443481) and a Mary K. Upson Visiting Professorship from Cornell University. Francisco Zamora-Arroyo was on sabbatical at the University of Arizona during the preparation of this article.

Notes

1. O. Hinojosa-Huerta, H. Iturribarría-Rojas, Y. Carrillo-Guerrero, M. de la Garza-Treviño, and E. Zamora-Hernández, *Bird Conservation Plan for the Colorado River Delta* (2004), available at http://www.sonoranjv.org/planning/deltabcp/BCP ColoradoDelta.pdf. The number of species in February 2009 was 371, compared to 368 reported in 2004; personal communication to the authors from Osvel Hinojosa-Huerta, February 2009.

2. M. Jenkins, "The efficiency paradox," *High Country News*, February 5, 2007, 8–13.

3. International Boundary and Water Commission (IBWC), "U.S. and Mexico meet in Phoenix, Arizona to address cooperative actions for the Colorado River basin," press release, March 13, 2008, available at http://www.ibwc.state.gov/Files/ PressRelease_031308.pdf. NGOs participating in the CRJCP include the Environmental Defense Fund, Pronatura Noroeste, and the Sonoran Institute.

4. O. Hinojosa-Huerta, J. García-Hernández, Y. Carrillo-Guerrero, and E. Zamora-Hernández, "Hovering over the Alto Golfo: The status and conservation of birds from the Río Colorado to the Gran Desierto," in *Dry Borders: Great Natural*

Reserves of the Sonoran Desert, edited by R. S. Felger and B. Broyles, 383–407 (Salt Lake City: University of Utah Press, 2007).

5. Hinojosa-Huerta et al., *Bird Conservation Plan for the Colorado River Delta.*

6. O. Hinojosa-Huerta, E. Iturribarría-Rojas, A. Calvo-Fonseca, J. Butrón-Mendez, and J. Butrón-Rodríguez, *Caracterización de la avifauna de los humedales de la Mesa de Andrade, Baja California, México*, Pronatura Noroeste report to the Instituto Nacional de Ecología (Mexico City: Instituto Nacional de Ecología, 2004).

7. G. García Saille, A. López López, and J. A. Navarro Urbina, "Lining the All-American Canal: Its impacts on aquifer water quality and crop yield in Mexicali Valley," in *The U.S.–Mexican Border Environment—Lining the All-American Canal: Competition or Cooperation for the Water in the U.S.–Mexican Border?* edited by V. Sánchez Munguía, 77–100 (Tijuana, Mexico, and San Diego: El Colegio de la Frontera Norte and San Diego State University Press, 2006).

8. F. Zamora-Arroyo, O. Hinojosa-Huerta, E. Santiago, E. Brott, and P. Culp, "Collaboration in Mexico: Renewed hope for the Colorado River delta," *Nevada Law Journal* 8(3) (2008): 871–89.

9. F. Zamora-Arroyo, J. Pitt, S. Cornelius, E. Glenn, O. Hinojosa-Huerta, M. Moreno, J. Garcia, P. Nagler, M. de la Garza, and I. Parra, *Conservation Priorities in the Colorado River Delta, Mexico and the United States* (Tucson, Ariz.: Sonoran Institute, Environmental Defense Fund, University of Arizona, Pronatura Noroeste Dirección de Conservación Sonora, Centro de Investigación en Alimentación y Desarrollo, and World Wildlife Fund–Gulf of California Program, 2005), available at http://sonoran.org/programs/ sonoran_desert/si_sdep_delta_priorities.html#report. Fifty-five experts from Mexico and the United States, including scientists, NGOs, and water stakeholders, identified the priority conservation areas.

10. Zamora-Arroyo et al., "Collaboration in Mexico."

11. IBWC, "U.S. and Mexico meet in Phoenix."

12. Morelos Dam is a diversion dam and provides no storage capacity. Approximately 90 percent of Mexico's allocation from the Colorado River is delivered by the United States at Morelos Dam.

13. U.S. Bureau of Reclamation, *Environmental Assessment for the Lower Colorado River Drop 2 Storage Reservoir Project Imperial County, California* (Washington, D.C.: U.S. Bureau of Reclamation, 2006), available at http://www .usbr.gov/lc/yuma/environmental_docs/Drop_2/draftea.pdf, last viewed December 4, 2006. The reservoir will have a capacity of 9.9 mcm (8,000 acre-feet).

14. C. W. Stockton and G. C. Jacoby, *Long-Term Surface and Streamflow Trends in the Upper Colorado River Basin*, Lake Powell Research Project Bulletin no. 18 (Washington, D.C.: National Science Foundation, 1976).

15. N. S. Christensen, A. W. Wood, N. Voisin, D. P. Lettenmaier, and R. N. Palmer, "Effects of climate change on the hydrology and water resources of the Colorado River basin," *Climate Change* 62 (2004): 337–63.

16. To secure the 51,000 acre-feet needed for a base flow (absent the dedication of water from other sources), NGOs will need to acquire water rights associated with 6,000 hectares (14,800 acres), or approximately 3 percent of the farmland in the irrigation district. This estimate for a base flow assumes that undiverted Colorado River water, operational releases, and current groundwater levels are maintained.

17. Comisión Nacional del Agua, *Actualización del estudio de factibilidad para la rehabilitación y modernización del distrito de riego 014, Río Colorado, B.C. y Sonora*, vol. 1 (Mexicali, Mexico: Informe General Organismo de Cuenca Península de Baja California, Comisión Nacional del Agua, 2006).

18. Yuma Desalting Plant and Ciénega de Santa Clara Working Group, *Balancing Water Needs on the Lower Colorado River* (2005), available at http://ag.arizona.edu/AZWATER/publications/YDP%20report%20042205.pdf, last viewed December 4, 2006.

19. Comisión Nacional del Agua, *Estudio de factibilidad ambiental, económica, técnica y social para el aprovechamiento de los escurrimientos superficiales enviados actualmente a los Estados Unidos de America a través del Río Nuevo* (Mexicali, Mexico: Organismo de Cuenca Península de Baja California, Comisión Nacional del Agua, 2007).

20. "Agreement among the State of Baja California, the Comisión Nacional del Agua, Pronatura Noroeste, and the Asociación Ecológica de Usuarios del Río Hardy y Colorado," 2007.

21. Imperial Irrigation District, *All American Canal Lining Project* (2006), available at http://www.iid.com/Water_Index.php?pid=64, last viewed December 4, 2006.

22. M. J. Cohen and C. Henges-Jeck, *Missing Water: The Uses and Flows of Water in the Colorado Delta Region* (Seattle: Pacific Institute, 2001).

23. Zamora-Arroyo et al., "Collaboration in Mexico."

Chapter 3

Restoring a Desert Lifeline

The Big Bend Reach of the Rio Grande

Mark Briggs, Carlos Sifuentes, Joe Sirotnak, and Mark Lockwood

In a Nutshell

- The Big Bend and Forgotten reaches of the Rio Grande support an important binational, desert-adapted ecosystem.
- Water impoundment, water withdrawal, and the spread of nonnative species have degraded ecologic function and structure; eighteen species of native fish have been entirely extirpated or are currently threatened.
- Small-scale restoration efforts must be linked with mid- and long-term strategic conservation planning that includes environmental water-rights acquisition and altered water releases from Río Conchos impoundments.
- Nongovernmental organizations can facilitate the binational, cooperative conservation necessary to restore and maintain the river system.
- Local communities, critical federal or local agencies, and the International Water and Boundary Commission should be included on the ground floor of any effort.

Introduction

For almost a decade, the World Wildlife Fund (also called the World Wide Fund for Nature, WWF), Big Bend National Park, Comisión Nacional de Áreas Naturales Protegidas (CONANP, National Commission of Protected Natural Areas), Big Bend Ranch State Park, and more than twenty other agencies, institutions, and organizations from both sides of the U.S.–Mexico border have been conducting a variety of activities to restore the Rio Grande's Big Bend reach.[1] Binational collaboration, the participation of divergent disciplines, and the involvement of riverside human communities are key ingredients needed to move forward to address conservation issues.

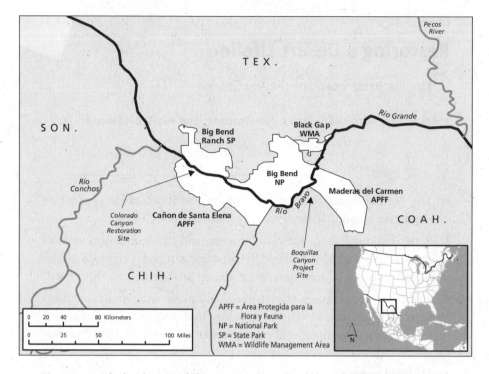

Figure 3.1. Idealized map of the Big Bend reach of the Rio Grande/Río Bravo showing federal and state protected areas involved in this binational conservation effort. At 3,000 kilometers (1,865 miles) long, the Rio Grande is the fifth-longest river in North America, flowing from headwaters in the southern Rocky Mountains of Colorado to a terminus near Brownsville, Texas. Its watershed occupies 46,700,000 hectares (180,000 square miles) of Mexico and the United States, including portions of Colorado, New Mexico, and Texas. The major tributaries of the Rio Grande are the Pecos River on the U.S. side and the Río Conchos in Mexico. Map drawn by Mickey Reed.

At El Paso and Ciudad Juárez, the Rio Grande/Río Bravo becomes the border between the United States and Mexico, and for some 1,995 kilometers (1,240 miles) it defines the southern limit of Texas and the northern limits of the states of Chihuahua, Coahuila, Nuevo León, and Tamaulipas before reaching the Gulf of Mexico near the cities of Brownsville and Matamoros (see fig. 3.1). The Rio Grande's Big Bend reach—defined for purposes of this chapter as the reach of the Rio Grande/Río Bravo between El Presidio, Texas, and Boquillas Canyon—winds through country dominated by canyons, mesas, ancient volcanoes, and calderas. The country that surrounds the Big Bend reach is remote, rugged, and beautiful, and floating the Rio Grande through Big Bend remains a world-class experience that ranks high on many boaters' "must-do" list.

Along this reach of the Rio Grande, the implementation of sound scientific investigations and rehabilitation projects, as well as water-management and wildlife habitat–conservation activities requires cross-border collaboration on a broad scale. Cross-border projects are often characterized by a variety of challenges that typically revolve around the flow of personnel, equipment, materials, and funding. Along the Big Bend reach, the inherent complexities of cross-border conservation have been increased by the aftermath of the attacks of September 11, 2001, and by drug smuggling and illegal migration, which together have made cross-border communication and cooperation more challenging.

As both a boundary river and an important river ecosystem, the Rio Grande carries significant international importance, supporting an exceptionally high diversity of wildlife. Its riparian woodlands are particularly important to conservation as a keystone habitat that exerts a powerful influence on biodiversity. Investigations have demonstrated that riparian ecosystems of the Southwest are some of the most productive ecosystems (defined by the biomass of organisms at each trophic level) in North America.[2] This point is underscored by the diversity of wildlife found along this important corridor. Some 351 bird species have been recorded in this region, and the corridor fulfills a large-scale ecological function as a major highway for birds and other species that migrate between the Americas.[3] In addition, the river supports twenty-three native species of fish.[4] For these and other reasons, the conservation of the Rio Grande's bottomland ecosystems has become a focus of binational conservation efforts.

Background and Discussion

A River in Decline and Its Drivers of Change

In contrast to the Rio Grande's captivating surroundings, its ecological condition has deteriorated significantly in recent years. River impoundment, overallocation of water, and the spread of nonnative plants and animals have taken their toll. Along various stretches, the river channel has become narrow and choked with invasive plants, just one manifestation of the river's changed hydrology. Altered physical and plant conditions along the river have negatively affected habitat for hundreds of species of wildlife.[5]

In conjunction with a set of biophysical changes that have taken place along the lower Rio Grande, numerous plant and animal species have become extinct or extirpated from the region, and many more have been listed on federal and state threatened and endangered lists. For example, seven native fish species are completely extirpated from the entire river, and eight other native fish have been extirpated from at least part of their historic range.[6] Of the twenty-three native fish species that remain in the region, eleven are listed as threatened or of conservation concern.[7] Due to these and other negative impacts, the Rio Grande/Río Bravo has been nominated four times as one of America's most endangered rivers.[8]

The Rio Grande basin is home to more than ten million people and includes some of the fastest-growing communities in the United States and Mexico. One source predicts that the combined Las Cruces–El Paso–Ciudad Juárez region may reach six million people by 2025, up from a current two million.[9] As agriculture, industry, and towns grow in the Chihuahuan Desert, the demand for water is mushrooming on both sides of the international border. River diversion for agricultural irrigation is by far the largest use of water throughout the Texas-Mexico portion of the Rio Grande basin, drawing between 67 and 90 percent of the water.[10] After agriculture, the largest consumer of Rio Grande water is evaporation from reservoirs and other open water surfaces. For example, the reservoir behind Elephant Butte Dam and the Caballo reservoir downstream in Mexico lose as much water to evaporation as is consumed by the city of Albuquerque, New Mexico, each year.[11] After evaporation, the next two consumers of Rio Grande water are municipalities and industry.

Before dams and flow regulation were put in place on the river, periodic large floods would inundate significant portions of the river's bottomland environment, eroding some areas and depositing fresh, moist alluvium in others. The unregulated river's active bottomland environment was thus relatively wide and open, with native cottonwoods and willows populating the floodplains in a patchlike pattern that also left many areas scoured and free of vegetation.

However, the U.S. Bureau of Reclamation's construction of Elephant Butte Dam in 1916 marked the beginning of the current preoccupation with water management whose objective is singularly focused on satisfying human needs. After that, numerous reservoirs were built on the Rio Grande and its tributaries, and widespread well fields were drilled, tapping the basin's aquifers. The dams, although supplying storage, have greatly diminished downstream flow (fig. 3.2). In some areas, groundwater pumping has

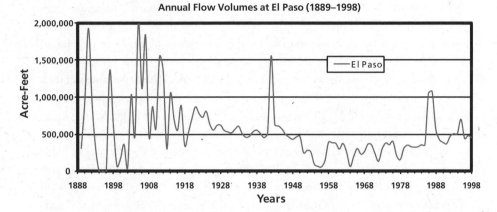

Figure 3.2. Total annual Rio Grande flow in acre-feet, measured at El Paso from 1889 to 1998. Note attenuation in flow following the construction of Elephant Butte in 1916.

reduced or even eliminated spring flow or allowed the infiltration of saline water into freshwater zones.

Prior to river impoundment, the Big Bend reach of the Rio Grande received much of its flow through the Forgotten reach—a 484-kilometer (300-mile) remote portion of the Rio Grande between El Paso and the river's confluence with Río Conchos. Today, however, flow that makes it through the Forgotten reach is significantly reduced owing to flow obstructions (sediment accumulation and dense salt cedar thickets) and increased loss from evapotranspiration and seepage, as well as impoundment and diversion upstream. Although some of these losses are offset by return flow and contributions from tributaries and springs along the Forgotten reach, annual-flow and peak-flow magnitudes are significantly reduced as compared to preimpoundment conditions.[12] Overall, flows in this peak period have been reduced as much as 77 percent.[13] In addition, the natural timing of flow through the Forgotten reach has been altered. Annual peak flows in the channel have shifted from May to July as a result of the storing and releasing of water, which are timed for irrigation purposes.

The hydrologic changes that have beset the Rio Grande have negatively affected the river's native bottomland plant communities. Floods scour and remobilize sediment for redeposition downstream. Freshly deposited, moist alluvium provides an ideal seedbed for many native bottomland plants, particularly when floods and consequent deposition occur during the spring, when many plants are producing seed. In short, flood disturbances are critical to the establishment of riparian vegetation, so the reduction in flood frequency and magnitude as well as the disruption of flood timing have created a "three-strikes-you're-out" scenario for many native bottomland plants.[14] As a result, the extent and distribution of such native bottomland plants as cottonwoods *(Populus fremontii)*, willow *(Salix goodingii)*, and ash *(Fraxinus velutina)* have been reduced to isolated small pockets, which in many cases are surrounded by dense stands of nonnative plants.

In contrast, several nonnative plants appear to be well-adapted indeed to the Rio Grande's altered hydrology. Topping the list in this regard are salt cedar *(Tamarix ramosissima, T. chinensis,* and *T. aphylla)*, which was introduced to the Rio Grande in the 1920s to address flood control and bank stabilization, and giant cane *(Arundo donax)*.[15] Both plants now dominate significant parts of the river's bottomland, occupying areas where native plants used to thrive or areas that were typically free of dense vegetation (e.g., areas prone to scour during preimpoundment times). Salt cedar produces seeds almost nine months of the year, and giant cane establishes rapidly and aggressively via rhizomes. Today, these two species form dense, sometimes impenetrable thickets along much of the lower Rio Grande. The combination of the establishment of dense stands of nonnatives on near-channel surfaces, reductions in stream flow due to river impoundment, stream-flow diversions, and episodes of drought has resulted in channel narrowing and vertical floodplain accretion that offer reduced-quality habitat conditions for many native species.[16]

Binational Restoration Efforts

Protected-area land managers, cooperating institutions, and other conservation agencies and organizations from both sides of the river are developing strategies to improve ecological conditions in a manner that is effective and affordable and that does not compromise the stakeholders who depend on the river for their livelihood and well-being. The protected areas themselves form the core of this binational effort. There are currently six protected areas along the Big Bend Rio Grande corridor: two in Mexico that are managed by CONANP (La Área Protegida de Cañon de Santa Elena and La Área Protegida de Maderas del Carmen) and three in the United States (Big Bend Ranch State Park, Big Bend National Park, and Black Gap Wildlife Management Area). In addition, 111 kilometers (69 miles) of the Rio Grande through Big Bend National Park have been designated as Wild and Scenic by the U.S. Congress, and two additional protected areas in Mexico,

El Monumento and Ocampo,[17] are poised to be designated during 2009, thus connecting and increasing protection of the river on the Mexican side between the two current protected areas. Together, the current suite of protected lands encompasses an area of 968,043 hectares (3,737 square miles) and protects roughly 225 kilometers (140 miles) of Rio Grande shoreline (fig. 3.1).

Over the past decade, a series of binational consultations and projects has been undertaken to respond to the Rio Grande's ecological decline. These binational efforts have been conducted with support and encouragement from organizations such as the WWF, the Rio Grande Institute (a local nongovernmental organization [NGO]), federal agencies (e.g., U.S. Geological Survey, the International Water Boundary Commission/ Comisión Internacional de Límites y Aguas [IBWC/CILA]), and a broad-based regional coalition, and have resulted in the implementation of on-the-ground rehabilitation efforts, securing water rights for environmental flow, conducting scientific investigations, and continuing monitoring efforts.

Small-Scale Projects to Remove Invasives and Reintroduce Native Plants

In several sites along the Big Bend reach of the Rio Grande, protected-area managers and WWF are implementing on-the-ground efforts to remove dense stands of nonnative plants and to reestablish native plants and natural channel conditions (fig. 3.3). The objective of these efforts is twofold: to reduce the extent and distribution of nonnative plants and to increase the vulnerability of near-channel sediments to erosion by current stream-flow conditions. Salt cedar and giant cane not only protect floodplain alluvium from erosion but also act as a perfect sediment filter, reducing flow velocities and flow energies, and thereby increasing aggradation.

The end result is that near-channel surfaces are becoming more permanent and built up. The morphologic trend of the channel through much of the Big Bend reach is thus narrower, deeper, and laterally stable, with accompanying loss of backwater and shallow areas that provide habitat for native fish and other species of wildlife. By aggressively attacking these dense nonnative stands, groups and institutions involved in this effort hope to precipitate a cascade of positive effects: removing nonnative plants will increase the vulnerability of underlying alluvium to mobilization, leading to an evacuation of sediment and a trend toward channel conditions that are wider, shallower, and laterally unstable. Such conditions are desirable because they increase channel conveyance, increase active channel surface

Figure 3.3. During the period from 2003 to 2006, binational work teams effectively treated more than fifty thousand salt cedar *(Tamarix ramosissima)* stems with handheld tools along the Boquillas Canyon reach of the Rio Grande. Photograph by Joe Sirotnak, Big Bend National Park.

area exposed to frequent inundation, and provide improved habitat conditions for a variety of native species (e.g., the federally listed and recently reintroduced Rio Grande silvery minnow, *Hybognathus argyritis*).

Several key insights have emerged from monitoring the results of these activities, with possible implications for similar transboundary conservation efforts:

1. Small-scale on-the-ground rehabilitation efforts alone produce limited ecological benefits.

The implementation of the on-the-ground efforts underscores an obvious realization: the entire Big Bend reach of the Rio Grande cannot be restored by these rehabilitation efforts alone. For one, it is impractical. Resource limitations (money, personnel, materials) combined with such practical considerations as access and the remote characteristics of the entire Big Bend reach itself limit the amount of on-the-ground work that can realistically be accomplished.

A second limitation is that small-scale, on-the-ground rehabilitation efforts typically do not address the root causes of ecological decline, but rather the symptoms. The explosion of salt cedar and giant cane along the Rio Grande is ultimately due to changes in stream flow brought on by impoundment, flow diversion, and myriad other actions and impacts that have made the river's hydrologic characteristics prone to ecological degradation. Therefore, efforts that focus solely on the eradication of nonnative plants miss the greater need to improve stream-flow conditions to support native bottomland habitat.

Eradicating salt cedar and giant cane along a specific river reach may bring ecological improvement to that specific part of the river but will probably not bring much benefit to reaches upstream or downstream. In contrast, altering or augmenting stream flow (e.g., via an environmental-flow program) or both have the potential to initiate a cascade of positive benefits felt through a large part of the riverine ecosystem (terrestrial as well as aquatic). For example, changing flow from the Río Conchos in Mexico to better reflect the needs of native bottomland habitat (while simultaneously meeting stakeholder and international treaty obligations) may bring strong ecological and social benefits to a large portion of the Big Bend reach. That said, implementing only environmental-flow programs (i.e., without on-the-ground rehabilitation efforts) will probably not be sufficient to remove salt cedar or giant cane. Therefore, reality dictates the implementation of both on-the-ground efforts and larger-scale flow management.

2. Small efforts can pave the way for much larger efforts.

It is important to note that the small-scale, on-the-ground efforts, despite their limitations, produce several useful and beneficial results. For example, the salt cedar–eradication effort in Boquillas Canyon was the first binational on-the-ground collaborative effort in the Big Bend area that directly involved personnel and materials from both sides of the border. As the first, its implementation allowed all participants to affect and improve upon the suite of activities required to conduct the project, such as securing binational work permits, improving communication, developing a binational payment process, transferring equipment from one side of the river to the other, and so on.

Of equal importance, on-the-ground efforts produce tangible results in a relatively brief time period. Having results to demonstrate is critical to creating momentum to conduct additional projects and implement complementary strategies that can affect larger areas (e.g., an environmental-flow program). In addition, having tangible results provides a clear track record that is invaluable for fund-raising.

3. Simpler design is often better.

Simple design and implementation strategies can be most effective toward achieving long-term project objectives. For example, two of the four rehabilitation projects implemented along the Big Bend reach involved the installation of expensive and labor-intensive irrigation systems to water native seedlings during the first two summers following planting. Instead, future projects may rely on natural regeneration or planting or both where natural moisture availability is adequate for plant establishment.

In contrast to the two rehabilitation projects that used irrigation, the other two efforts were carried out with relatively meager resources. For example, the effort to remove salt cedar from Boquillas Canyon was accomplished largely with personnel from Maderas del Carmen in Mexico and labor from the nearby town of Boquillas, Mexico. Eradication methods relied on removal strategies involving simple hand tools and spot herbicide application with hand-held bottle sprayers. Since the inception of this effort in 2003, more than thirty thousand salt cedar stems have been treated, with post-treatment monitoring data revealing high and impressive mortality rates.

4. Consider labor costs and participation benefits.

Although the treatment of salt cedar has been carried out in Boquillas Canyon only once a year, the work has provided economic benefits (a two-week salary) to more than twenty citizens of the town of Boquillas each year for the past four years. This employment, in turn, has sparked community interest and an improved understanding of conservation work along the river. These efforts also force agencies and organizations from both sides of the border to work together. Resolving coordination challenges benefits each of these small-scale efforts and is a prerequisite for addressing such long-term challenges as river management. Future rehabilitation efforts along the Big Bend reach of the river will be broader in scope, providing additional employment opportunities for riverside citizens.

Project to Secure Water Rights in the Forgotten Reach

Environmental Defense Fund, WWF, the Trans-Pecos Water Trust, and a new NGO soon to be established in Mexico are pursuing the strategy of acquiring water rights in the Forgotten reach to support the long-term viability of bottomland habitat along the Big Bend reach of the Rio Grande and the Rio Conchos. The river's bottomland ecosystem would become another recognized water user or stakeholder, with environmental-flow rights supporting the establishment and proliferation of native species (e.g., augmenting flow during times of drought and in periods critical for the proliferation of native species).

If successful, the environmental-flow program would provide clear long-term benefits for the Rio Grande's bottomland ecosystem, addressing at least some of the root causes of ecological decline. Although there are significant political, legal, and ecologic hurdles today, initial research has demonstrated that environmental water rights are available and that obtaining these rights, although expensive, is probably a viable option.

Of potentially greater importance than securing additional rights is altering the management of dams and the timing of water releases to reflect the river's ecological needs. In this regard, the greater opportunity may be along the Río Conchos in Mexico. The Río Conchos has the strongest influence on flow characteristics along the Big Bend reach of the Rio Grande, and its flow is managed under the 1944 International Water Treaty between Mexico and the United States.

Stakeholders including WWF, CONANP, the Comisión Nacional del Agua (National Water Commission), and IBWC/CILA are investigating how releases from the Conchos can be modified to improve river conditions for riverside citizens and bottomland native habitat without violating treaty obligations. Targeting releases seasonally to provide for key riverine biologic needs and altering discharge magnitudes during the release period (to provide for a more natural hydrograph) are just two examples of such modifications.

However the environmental flow program manifests itself, its overall effectiveness will be enhanced by on-the-ground restoration activities to reduce nonnative species and to enhance the proliferation of native species or to do both. Simply providing greater amounts of water would likely support the exotic invasive species that have already established along the river (e.g., more giant cane and salt cedar). In the Forgotten reach, future efforts may focus on the temporary ponds that form when flow is forced out of the channel, potentially manipulating their vegetation and morphologic conditions to enhance their habitat quality for migratory birds and other wildlife.

Conclusions

Despite common ground established among protected-area managers, the development of a true binational program along the Big Bend reach of the Rio Grande has not been easy and remains a work in progress. In addition to challenges inherent in working across an international boundary, the protected areas of Big Bend occupy a huge remote and rugged region, with significant travel time required to move between sites. In addition, the number of official river crossings has declined in recent years, significantly adding to the travel time for managers to visit other managers across the border. Given the significant deterioration of the river's ecology and the sociopolitical hurdles that must be overcome to address the river's ecological decline, restoring the Big Bend reach to predam conditions may not be possible.

Nevertheless, there remains considerable incentive to continue binational collaborative projects owing to the shared common resource that is encompassed by the six Big Bend protected areas and the benefits that improving the river's ecological conditions will bring to native species, riverside human communities, visitors, and other stakeholders.

Despite a recent record of U.S.–Mexico political tensions, significant communication and collaboration have occurred between natural-resource managers on both sides of the river.

NGO participation has also proved critical for organizing initial discussions among stakeholders, providing the spark for developing the binational foundation on which the rest of the overall project can be constructed. In the Big Bend situation, it is difficult for agency personnel to reach outside the boundaries of their own protected area, much less across an international boundary. In the years prior to the implementation of the binational work along the Rio Grande, WWF and Rio Grande Institute staff traveled between protected areas and land managers to promote communication and cooperation. NGO participation has also provided additional avenues for fostering participation of local communities, creating jobs, raising funds, and providing a means to distribute funds to areas outside the agency.

There are currently no set policies or incentives to encourage land managers on either side of the river to contact and work closely with their international counterparts. Binational efforts would greatly benefit from a change in federal and state policies toward the promotion of collaborative cross-border activities in a manner that encourages and rewards natural-resource managers who reach across their borders to participate in such activities.

Of course, fund-raising must support the binational efforts highlighted here. In addition to the challenges inherent in raising funds, secured funding often comes with restrictions (e.g., the money can be used only on one side of the river), an obvious challenge for cross-border projects. We strongly recommend that potential funders of binational collaborative projects offer support to all stages of the collaborative conservation project. These stages include not only implementation activities (on-the-ground rehabilitation efforts), but also preimplementation activities (organizational meetings, natural-resource evaluation, data gathering and assessment, and so on) and postimplementation activities (monitoring and evaluation). Such overall support is an absolute prerequisite to developing binational collaboration because it provides the basis for communication, organization, and planning.

The IBWC/CILA mission is to provide binational solutions to issues that arise during the application of U.S.–Mexico treaties regarding boundary demarcation, national ownership of waters, sanitation, water quality, and flood control in the border region.[18] The commission's knowledge and support of the efforts along the Rio Grande from the beginning proved critical from both a legal standpoint (they needed to know what was happening along the U.S.–Mexico border) and a practical standpoint (e.g., securing short-term permits that allowed U.S. and Mexican participants to cross the river in designated project areas).

In addition, involving residents from riverside communities is a critical component with numerous benefits. The town of Boquillas, Mexico, lies upstream of one of our project areas, and, as noted, more than twenty of its citizens have been trained and hired to treat salt cedar on both sides of the river (with temporary work visas secured under IBWC/CILA auspices). Residents from Boquillas and the town of El Presidio have been hired to maintain two pilot restoration efforts, assisting agency personnel in weeding, repairing irrigation lines, and monitoring. Such community participation provides benefits to both the projects (a local and willing source of labor) and residents (paid employment). Providing direct economic support to local residents needs to remain a priority in this and other conservation efforts.

Finally, all stream-rehabilitation efforts—binational or not—need to address at some level the root causes of ecological decline, regardless of known political, legal, and monetary hurdles. Only by doing so can long-term ecological improvement be realized.[19] Ecological decline along the Big Bend reach of the Rio Grande is due largely to changes in hydrologic conditions brought about by river impoundment and flow diversion. Restoration can be addressed in the long term only by altering river management, acquiring environmental flow, and changing releases from the Río Conchos. Although there is a long way to go before significant headway can be made in improving the ecological conditions of this portion of the Rio Grande, these preliminary collaborative, binational experiences have provided a strong foundation from which additional efforts can be launched in the future. It may be impossible to return the river to its predam ecological condition in our lifetimes, yet the progress made so far can lead to real ecological improvement that will benefit both the river environment and stakeholders.

Notes

1. See vision statement and results of 2008 Rio Grande binational workshop on WWF's Chihuahuan Desert Web site: http://www.worldwildlife.org/what/wherewework/chihuahuandesert/index.html.

2. R. R. Johnson and D. A. Jones, tech. coords., *Strategies for Protection and Management of Floodplain Wetlands and Other Riparian Ecosystems. Riparian Habitat: A Symposium, July 9, 1977, Tucson, Arizona*, Forest Service General Technical Report RM-43, U.S. Rocky Mountain and Range Experiment Station (Fort Collins, Colo.: Forest Service, U.S. Department of Agriculture, 1977); R. R. Johnson and J. F. McCormick, tech. coords., *Strategies for Protection and Management of Floodplain Wetlands and Other Riparian Ecosystems*, symposium proceedings, Forest Service General Technical Report WO-12 (Washington, D.C.: Forest Service, U.S. Department of Agriculture, 1978).

3. N. G. Stotz, *Historic Reconstruction of the Ecology of the Rio Grande/Rio Bravo Channel and Floodplain in the Chihuahuan Desert*, report (Las Cruces, N.Mex.: Chihuahuan Desert Program, World Wildlife Fund, 2000).

4. C. Hubbs, R. R. Miller, R. J. Edwards, K. W. Thompson, E. Marsh, G. P. Garrett, G. L. Powell, D. J. Morris, and R. W. Zerr, "Fishes inhabiting the Rio Grande, Texas and Mexico, between El Paso and the Pecos confluence," in *Importance, Preservation, and Management of Riparian Habitat: A Symposium*, 91–97, Forest Service General Technical Report RM-43 (Washington, D.C.: Forest Service, U.S. Department of Agriculture, 1977).

5. Stotz, *Historic Reconstruction of the Ecology of the Rio Grande/Rio Bravo Channel*; American Rivers, *America's Most Endangered Rivers of 2000* (Washington, D.C.: American Rivers, 2000).

6. Stotz, *Historic Reconstruction of the Ecology of the Rio Grande/Rio Bravo Channel*.

7. Ibid.

8. American Rivers, *America's Most Endangered Rivers of 2000*.

9. Stotz, *Historic Reconstruction of the Ecology of the Rio Grande/Rio Bravo Channel*.

10. M. E. Kelly, *The Río Conchos: A Preliminary Overview* (Austin: Texas Center for Policy Studies, January 2001).

11. P. Gleick, "Making every drop count," *Scientific American* (February 2001): 40–45.

12. D. J. Dean and J. C. Schmidt, "From channel to floodplain: Historic geomorphic transformation of the lower Rio Grande in the Big Bend region of Texas, Chihuahua, and Coahuila," in Multi-scale Feedbacks in Ecogeomorphology, special issue of Geomorphology (in review).

13. M. E. Landis, *The Mighty Río Grande, Fort Quitman–Presidio: The Sometimes Gone and Oft Forgotten River* (El Paso, Tex.: El Paso Field Division, U.S. Bureau of Reclamation, 2001).

14. C. R. Hupp and W. R. Osterkamp, "Bottomland vegetation distribution along Passage Creek, Virginia, in relation to fluvial landforms," *Ecology* 66 (1985): 670–81; C. J. Campbell and W. Green, "Perpetual succession of stream-channel vegetation in a semiarid region," *Journal of the Arizona Academy of Science* 5 (1986): 86–97; B. L. Everitt, "Use of the cottonwood in an investigation of the recent history of a flood plain," *Science* 266 (1968): 417–39.

15. C. S. Crawford, A. C. Cully, R. Leutheuser, M. S. Sifuentes, L. H. White, and J. P. Wilber, *Middle Rio Grande Ecosystem: Bosque Biological Management Plan* (Albuquerque: U.S. Fish and Wildlife Service, 1993).

16. D. J. Dean, "A river transformed: Historic geomorphic changes of the lower Rio Grande in the Big Bend region of Texas, Chihuahua, and Coahuila," master's thesis, Utah State University, Logan, 2009; Dean and Schmidt, "From channel to floodplain."

17. During the final stages of editing this chapter, the Mexican government made the official declaration of Área Natural Protegida de Flora y Fauna del Ocampo, which protects an additional 344,238 hectares of the Big Bend/El Carmen complex.

18. See the International Boundary and Water Commission official Web site at http://www.ibwc.state.gov/home.html, accessed December 2007.

19. M. K. Briggs, *Riparian Ecosystem Recovery: Strategies and References* (Tucson: University of Arizona Press, 1996).

A Convergence of Borders

Indigenous Peoples and Environmental Conservation at the U.S.–Mexico Border

Rachel Rose Starks and Adrian Quijada-Mascareñas

In a Nutshell

- Close to forty Indigenous communities, Native nations, and traditional land-use areas are located near to the current international border; many have traditional connections across that border.
- In Mexico, no formal legal recognition exists for autonomous Indigenous communities. This situation constitutes one of the major challenges to involving Indigenous communities in border conservation, environmental regulation, and policy.
- In the United States, the system of federal recognition provides a legal framework through which Indigenous governments are formally represented. This recognition, however, provides little opportunity for including Indigenous governments in environmental regulation or conservation policymaking.
- Despite these legal barriers, effective partnerships for implementing conservation policy have been created through voluntary programs initiated by federal governments, Native communities, and other organizations.
- Progress requires a systematic legal framework in which Indigenous communities and their governments can have access to the governmental processes that drive environmental regulation and conservation policy.

Introduction

The international border between Mexico and the United States separates more than two neighboring countries. This political line also separates specific Indigenous lands and people, and in the process it artificially divides ecosystems and landscapes.

In the United States, a formalized system based on a complicated history between tribes and the United States creates a government-to-government relationship between sovereign Native nations and the U.S. federal government. The reservation system provides for a form of territorial integrity but does not generally extend Native jurisdiction to the full extent of traditional use or concern. This legal framework provides few opportunities for addressing conservation issues. In Mexico, no such system of governmental recognition exists. Many indigenous communities within Mexico live in *ejidos*, nonautonomous rural communities with traditions of local government that are not formally recognized by either state governments or the federal government.

Indigenous communities in close proximity to the U.S.–Mexico border face the same concerns as their neighbors, including vegetation loss and contamination, water quality and quantity, and the impacts of migration and industrial development. Moreover, they face these concerns across not one boundary, but *more than thirty* international and intertribal boundaries, and they face them in an environment complicated by issues of sovereignty, jurisdiction, and institutional process.

Indigenous communities and sovereign Indigenous nations approach conservation in the context of their historical and contemporary relationship with the land, based on spiritual and cultural practices, traditional subsistence and interaction with plants and animals, and an ongoing responsibility to care for their land. In addition, as they tackle contemporary environmental concerns in the border region, they work across jurisdictions with state, national, and international agencies and organizations. Native communities are thus critical stakeholders in border conservation, but their role should be expanded. Along with greater funding for infrastructure and capacity building, Indigenous communities and their governments require meaningful access to the institutional processes that lead to environmental regulation and policy.

This chapter provides a description of the unique legal status of Indigenous people and lands in the border region as it affects their ability to address their environmental concerns. It surveys some of the activities that Native communities in both countries are pursuing to better manage and conserve the area's natural resources, with examples in three key areas: water, plants, and migration impacts. It concludes with a set of recommendations to enhance the involvement of Indigenous stakeholders in conservation policymaking in the U.S.–Mexico border context.

Native Nations at the U.S.–Mexico Border

Indigenous nations and communities in the southwestern United States and northern Mexico (see table 4.1) have a rich history of culture, arts, agriculture, environmental management, and governance that predates the current conversation on the border.[1] To many Native cultures, the U.S.–Mexico border is an artificial line imposed by outside governments and should have little influence on Native life. However, current policies and impacts of security, immigration, and environmental degradation have brought the border to the forefront among Native concerns. In addition, the maintenance of Native culture and familial ties across the border is often interrupted or severed by U.S. or Mexican policies that impede the physical crossing of or communication across the contemporary political boundary line.[2] There is no way to measure fully how much this situation negatively impacts the capacity of tribal members and tribal governments from both sides of the border to act in concert around environmental or social issues.

Indigenous nations (United States) and communities (Mexico) have very different legal statuses within their respective countries, so recommendations for Indigenous involvement differ for the United States and Mexico. Whereas Native nations in the United States approach conservation policy as tribal governments, Mexican Indigenous communities lack this type of formal recognition and avenue for influencing the making and implementation of policy.

Status of Native Peoples and Lands in the United States

Within the United States, Native nations occupy a legal and historical status distinct from other groups. Since early contact with Europeans, Indigenous tribes have been regarded as sovereign political entities, with rights to control territory. This right of sovereignty—although contested, diminished, and partially extinguished through several centuries of U.S. legislation, executive orders, and court decisions—remains one of the U.S. Indigenous nations' most valuable tools in governance and, by extension, for conservation.[3]

In the United States, each Native nation with a reservation has a border that separates an area of tribal jurisdiction from state and other local jurisdictions. At the same time, traditional territory typically extends beyond the reservation boundaries and, in the case of U.S.–Mexico border tribes, beyond international boundaries. Reservations are federally recognized land

Table 4.1. Indigenous communities and nations in the U.S.–Mexico border area, by state.

—Indigenous Communities in Mexico—

Baja California: Cochimí, Cucapá, Kiliwa, Kumiai, Pai Pai

Coahuila: Kikapu

Sonora: Pápago, Yaqui

—Tribal Reservations in the United States—

Arizona: Cocopah, Fort Yuma–Quechan, Pascua Yaqui, Tohono O'odham

California: Barona, Campo, Capitán Grande, Cuyapaipe, Inaja and Cosmit, Jamul, La Posta, Manzanita, San Pasqual, Santa Ysabel, Sycuan, and Viejas (all Kumeyaay); Pala (Cupeño and Luiseño); La Jolla, Pauma and Yuima, Pechanga, and Rincon (all Luiseño); Los Coyotes and Torres-Martínez (both Cahuilla)

New Mexico: Mescalero Apache

Texas: Kickapoo, Ysleta del Sur (Tigua)

---

with discrete, jurisdictional boundaries, whereas traditional territory is land that Native nations have long used and considered to be part of their territory, but over which they may not have formal, jurisdictional control.

Like states, many tribes have governments with legislative, executive, and judicial branches, territory to govern, police forces, environmental agencies, and citizens who vote and make decisions through elected leaders.[4] With exceptions in certain states, their land, people, and policies do not fall under the jurisdiction of the U.S. state in which they are located.[5] Instead, tribal courts and tribal councils handle many legal questions; U.S. federal (not state) courts hear cases that are not within tribal jurisdiction. Unlike states, Native nations are not separately represented in the U.S. Congress, although tribal populations do participate in local, state, and national elections. In any negotiation with the federal government or with a state, Native nations come to the table as governments, not merely as minority or interest groups.

Although the nature and limits of tribal sovereignty are well defined vis-à-vis criminality laws (see Major Crimes Act of 1885, 18 U.S.C. §1153), no well-defined framework exists for conservation or environmental regulation.

That is, there is no similarly clear legal framework for tribal governments to participate in forming environmental regulations.

Status of Native Peoples and Lands in Mexico

Rather than a reservation system, Indigenous land tenure in Mexico is largely based in ejidos. A modern version of pre-Columbian land use, the ejido system includes parcels of land shared by a community's people. From the Spanish colonial period up to the twentieth century, the feudal *encomienda* system, which upheld social inequalities, had been the land-tenure policy.[6] After the Mexican Revolution (1910–20), the constitutional encomienda system was abolished; this action affirmed the right to communal land for traditional Indigenous and mestizo (mixed European and Amerindian ancestry) Mexican communities.

The ejido system was reintroduced as an important component of the land-reform program. However, in 1991, the Mexican government amended the Constitution's Article 27, the law that regulates land tenure and natural resources, and thereby opened the possibility to lease or sell communal land. Since then, some ejido land has been sold to corporations, although much of it is still in the hands of Indigenous and mestizo Mexican farmers.

Although Indigenous communities may hold communal land parcels, they do not have formal jurisdiction over their territory. Depending on the Mexican region, they often share land administration with mestizo Mexicans.[7] There is room, however, for autonomous organization of local administration for communication with governmental agencies. Most Indigenous Mexican communities are represented by a community governor, and decisions regarding social or environmental concerns are made communally through traditional councils. In some cases, as for the Pápagos in Sonora (Tohono O'odham in Arizona), the government structure resembles Mexican administrative divisions with executive, legislative, and judicial branches.[8]

Indigenous land rights within Mexican law are unevenly implemented due to inconsistencies in legislation. Perhaps the most comprehensive implementation of Indigenous rights is found in the Ley General del Equilibrio Ecológico y Protección al Medio Ambiente (General Law of Ecological Equilibrium and Environmental Protection). This law explicitly takes Indigenous communities into consideration in several areas of local and federal importance: environmental policy, protection of natural areas and wildlife, social policy, and environmental information.[9] It requires consultation of and participation from Indigenous communities. For instance, protected natural areas can be proposed, administered, and managed by Indigenous communities in conjunction with other local and federal agencies.

Despite the recent advances in Indigenous participation within Mexico, governmental capacities are still limited by structures that provide little more than consultative rights. Mexico's Indigenous communities have difficulty engaging in policymaking because they lack formal recognition and power, which means they have diminished ability to participate in conservation policymaking. Whereas the U.S. federal system lacks an organized method for Native nations' participation in environmental regulation, the Mexican system lacks a legal provision to recognize Indigenous sovereignty. Native governments in the United States do not currently have a seat at the environmental policymaking table, but it is clear who would participate if they did; in Mexico, there is no recognized system for determining who would occupy a seat at the table, even if they had one.

Case Studies in Indigenous Peoples and Conservation at the Border

Many Native peoples maintain a strong link between culture and the environment. They often consider the physical landscape sacred—the home of medicinal plants, ceremonies, and ancestral burial grounds, as well as a source of spiritual connection.

Habitat and ecosystem degradation is happening throughout the world, affecting many Native lands, but living in the border region brings special concerns. Near the U.S–Mexico border, Native communities' local ecosystems are detrimentally affected by practices on either side of the border and beyond their traditional territories.

For U.S. tribes, conservation becomes an international issue because they must address it in an international context. The following examples illustrate Native communities' diverse ways of working on conservation issues that focus on both culture and the maintenance of biodiversity in the border region. These cases focus on water, plant resources, and migration policy impacts.

Transboundary Water Conservation

The Colorado River transboundary habitat has been an area of focused conservation attention. Two tribes, the Cocopah (Cucapá in Mexico) and the Quechan, have joined with dozens of agencies and nongovernmental organizations (NGOs) to protect this fragile habitat.

The lower Colorado River is central to Cocopah life in southwestern Arizona, northern Sonora, and Baja California. That life has changed since the river has been dammed and diverted upstream.[10] The river has largely dried up, causing people to move away from it as a source of subsistence. The Arizona Cocopah population along the Colorado has dropped from about five thousand to eight hundred residents over the past generation.[11] To the south, the Cucapá have experienced river contamination from pollutants and toxic dumping that has occurred upstream within the United States.[12] The river, which today has drastically reduced flows (see chapter 2 in this volume), formerly supported Cocopah agriculture—including plantings in the floodplain, the use of wild rice, wheat, and other wild greens and grains, and the use of grasses for basketry—as well as hunting and fishing that sustained the tribe.[13]

In 2002, through the Colorado River International Conservation Area committee, the Cocopah Indian Tribe began working with the National Wildlife Federation and the U.S. Bureau of Land Management (BLM) to restore the river and its resources from the Cocopah Reservation to the river delta in Mexico. The Cocopah Indian Tribe and the National Wildlife Federation formed the committee to obtain resources to protect the habitats of the limitrophe (the 37-kilometer [23-mile] portion of the Colorado River connecting the two countries).

The BLM has further proposed coordinated management of the river on federal lands.[14] Other conservation efforts have received funding from various U.S. government agencies in addition to the BLM: the Environmental Protection Agency (EPA), Bureau of Indian Affairs, Bureau of Reclamation, Fish and Wildlife Service, Department of Homeland Security, and National Fish and Wildlife Foundation. The collective goal is to make the lower Colorado River an international conservation area.[15]

These ongoing efforts involve negotiating with various stakeholders, especially on water rights, allocations, and water distribution plans, as well as arguing to maintain as much water in the river as possible.

Also on the lower Colorado, the Quechan Indian Tribe worked with the City of Yuma, the State of Arizona, and the Yuma Crossing National Heritage Area on the Yuma East Wetlands Project to restore approximately 570 hectares (1,400 acres) along the river.[16] About half of the target area runs through the Quechan Reservation. This area, once a place for hunting and fishing, became overrun by invasive nonnative plants, which provided cover for illegal border activity.[17] The restoration partnership began in 2002 when the City of Yuma and the Quechan Indian Tribe agreed to fund the restoration of the Ocean-to-Ocean Bridge, spanning this portion of the Colorado River.[18] This wetland-restoration cooperative has won several awards, including Project of the Year from the National American Public Works Association.

To further this restoration, a binational meeting was held in April 2008 by the Yuma Crossing National Heritage Area and the Mexican NGO Pronatura Noroeste to announce a joint agreement to restore the limitrophe.[19] Although this particular agreement is a broader partnership of U.S. and Mexican organizations, the Quechan and Cocopah/Cucapá are important stakeholders in this habitat restoration.

These examples highlight the productivity of effective collaboration among tribes, NGOs, government agencies, and international partnerships around the border region's water resources. Because these tribes depend heavily on riparian habitats, they have much at stake in conservation and restoration initiatives and must continue to be partners in these projects.

Managing Culturally Significant Plant Resources

Many Native communities depend on indigenous vegetation resources for food, art, and ceremonial practices. For instance, the Tohono O'odham/ Pápago, use bear grass *(Xerophyllum tenax)*, yucca (*Yucca* spp.), and devil's claw (*Proboscidea* spp.) for basket weaving. Weavers say that the art of traditional weaving strengthens their voices, leadership, and traditions. But widespread use of pesticides and herbicides can both harm the natural plants and be toxic to the weavers who handle the material.[20]

On another front, the citizens' group Tohono O'odham Community Action works to change tribal citizens' eating habits from diets high in sugar and fat by renewing the use of foods native to the Sonoran Desert. These native foods include saguaro fruit *(Carnegiea gigantea)*, cholla cactus buds (*Opuntia* spp.), and mesquite beans (*Prosopis* spp.).[21]

The Kumeyaay/Kumiai in southern California and Baja California also utilize indigenous plants for their basketry and for food, clothing, housing, and medicine.[22] Native weavers in California have formed their own NGO to advocate for their art. Since 1988, the California Indian Basketweavers Association (CIBA) has been involved in trying to curb pesticide spraying throughout California's public-use areas, especially national forests.[23] Public-use areas are important to these efforts because weavers, especially those from California tribes with very small reservations, have little land to gather their plant materials, making public land an important resource for them. In partnership with the California Indian Forest and Fire Management

Council (CIFFMC), CIBA has negotiated with the U.S. Forest Service (Pacific Region) and California's BLM to create policies for gathering traditional materials and to protect access to the grasses and other plants needed for their art.[24]

In November 2006, the U.S. Forest Service Pacific Region and the California BLM signed an interagency policy on the use of culturally important plants found on the lands managed by the two agencies. This policy allows for permit-free traditional gathering by Native practitioners and promotes native plant sustainability through local consultation with tribes on "management practices to restore, enhance, and promote ecosystem health."[25]

This interagency agreement resulted from collaboration through the Gathering Policy Working Group. The group, aiming to create a policy that protects Native gathering practice, is made up of personnel from the two governmental agencies, CIBA and CIFFMC. It convened a series of listening sessions to obtain input from traditional gatherers and tribal leaders.[26] CIBA and CIFFMC have offered their experience to other tribal groups to create similar agreements in their regions. Although much of this work was done in central and northern California, it might be a model for forming national policy and useful for collaboration among cross-border agencies.[27] This case illustrates the linkages between conservation and Native traditions and provides a model for collaboration among NGOs, governmental agencies, and community members.

Mitigating the Impacts of Border Policy and Migration on Native Lands

When Native communities are split by the international border, many basic community connections are strained or severed. U.S. border enforcement has had enormous impacts on U.S. tribal lands and communities.[28] Beginning in 1993, border policy meant to curb undocumented immigration in urban areas has altered migrant routes into rural areas (see chapters 13, 14, and 15 in this volume), causing overwhelming traffic on some reservations (in particular the Tohono O'odham Reservation in Arizona).[29] The increased traffic has led to increased U.S. Border Patrol presence, environmental degradation, illegal trafficking, and disrupted Native communities. These impacts became more severe after the increase in security following the attacks of September 11, 2001.[30]

On Tohono O'odham land, some residents are afraid to go out into the desert to collect saguaro blossoms or to herd their sheep near the border. They have experienced harassment and fear violence resulting from increased militarization at the border.[31] Farther to the east, the Tigua of Ysleta del Sur Pueblo have objected to the fence to be raised along the border at El Paso. They will be cut off from sacred grounds along the Rio Grande where they have held celebrations and ceremonies for hundreds of years.[32]

On the Tohono O'odham Reservation, increased migrant and Border Patrol traffic has meant that as new official and unofficial roads are created, archaeological sites, burial grounds, and natural resources are damaged.[33] Patrol and migrant vehicles destroy plants and run over animals, such as the desert tortoise *(Gopherus agassizii)*.[34]

One of the greatest impacts immigration leaves on tribal lands is the increase in trash. The estimated half-million undocumented border cross-ers each year leave about two thousand tons of trash on land throughout southern Arizona,[35] about one hundred tons of which were removed from O'odham land in 2005 and 2006 alone.[36] The Tohono O'odham Solid Waste Management Program collected, sorted, and recorded the types of trash found on their land to help assess the options for removal and recycling. In 2004–2005, the nation sorted forty-five tons of solid waste, finding unus-able waste along with recyclable and reusable material such as backpacks, clothing, blankets, water bottles, plastic sheeting, and food.[37] In addition, in 2005, the nation found 956 bicycles that can be refurbished and 120 abandoned vehicles that can potentially be sold or used within the O'odham communities.[38]

To address the damages to Tohono O'odham land caused by border activity, the tribe has undertaken a significant waste cleanup effort. They have done so with cooperation and funding from the Arizona Department of Environmental Quality, the U.S. BLM, and the EPA through the BLM's Southern Arizona Project.[39]

Recommendations

The primary barrier to Indigenous stakeholders' further engagement in conservation planning and environmental regulation is structural. These stakeholders do not have access to the institutional processes that create policy; they do not have statutory authority to participate in regulation that affects their jurisdictions; and they do not have influence over the bounds of national and international regulatory authority. Thus, underlying each recommendation is the imperative to create a consistent legal framework in which Indigenous perspectives and concerns are addressed and under which Indigenous stakeholders have a clearly defined role in policymaking.

For example, although border tribes have significant concerns about the future of the region's watersheds and habitats, they are rarely invited to participate in forming the federal- and state-level policies in Mexico or the United States that directly impact the border environment. The examples given in this chapter describe situations in which agencies or Indigenous communities have made a special, voluntary effort to form partnerships, and they demonstrate that many Native communities and their leaders have the capacity and desire to be involved in natural-resources policy in the border region. Some tribes, despite their complex and constraining legal and historical context, are beginning to find ways to bridge jurisdictions and interests to address environmental and social issues related to the border.

Any solution to problems faced by Native nations in the border region necessarily involves many agencies and jurisdictions. Many U.S. agencies have written policies on tribal consultation. Although this official acknowledgment of tribes' interests is positive, being merely *consulted* on a federal project may not translate into meaningful access to the institutional processes that create environmental regulations. In response to this problem, the Indigenous Peoples Subcommittee of the National Environmental Justice Advisory Council suggests, among other things, the development of ongoing relationships with tribes, allowance of ample time for tribal consideration, and the creation of institutionalized policies and protocols to make the consultation process more meaningful.[40]

These policy suggestions provide starting points for involving Indigenous governments and communities more deeply in conservation decision making in the U.S.–Mexico border context.

Formalization of Representation in Institutional Processes, United States

Tribes are often involved very late in the policymaking process. Partnering with tribes includes involving them early in the decision-making process, as was done with the gathering policy for native plants in California public lands and the partnerships for lower Colorado River restoration.

For future collaboration, the Border Governors Conference is an example of a forum where tribal executives could represent border tribes' concerns. (The Border Governors Conference is an annual meeting of the governors of the four U.S. states and six Mexican states along the border. The conference agenda includes topics such as agriculture and livestock, border crossing, economic development, education, energy, environment, health, science and technology, security, tourism, water, and wildlife.) *Executives of the border tribes can be included in this conference as the leaders of sovereign governments.* In addition, agreements among agencies and tribes can be made through official statements, memoranda of understanding, policy and protocols, regular contact and relationship building, and official legislation requiring cooperation at the federal, state, and local levels.

Formalization of Representation in Institutional Processes, Mexico

For Indigenous communities in Mexico, the framework for participation is less clear. Investment in an Indigenous governance infrastructure may be the proper precursor to high-level participation and representation; the timescale required for institutional development may necessitate interim measures—for example, providing for the election of Indigenous Mexican representatives to the Border Governors Conference from an intertribal electorate.

Funding and Expansion of Border 2012

The U.S. EPA's Border 2012 Program has been a useful source of funding and other resources to tribes. In 2005, Indigenous leaders were included for the first time in the National Coordinators Committee meeting. The Arizona and California Tribal Liaisons meet yearly with Indigenous leaders in a tribal caucus to focus on issues affecting tribes. The topics of this caucus are published regularly in a report to the National Coordinators Committee. *This type of program needs to be continued, while the development of more meaningful dialogue between Native leaders and the Border 2012 National Coordinators Committee is pursued.* Tribal leaders should be members of the National Coordinators Committee, with tribes nominating or choosing representatives from their caucus to fill one or more seats at the committee table.

Continued Capacity Building

Many tribes have developed expertise in environmental management. Examples include the Tohono O'odham migrant-waste cleanup; restoration along the Colorado River by a partnership of agencies, organizations, and the Cocopah and the Quechan tribes; and the traditional plant materials policy developed in part by the Gathering Policy Working Group in California. *These partnerships were effective because the Native partners were involved in creating policy rather than being informed of the policy after it had already been made. More binational cooperation is needed with other agencies and among tribes; funding for infrastructure and capacity building is also needed.* Priority funding for infrastructure would support the development of databases for conserving and managing Native lands, as well as the creation of educational programs to increase environmental awareness. For example, the NGO Pronatura designed a new initiative recently to strengthen legal and institutional tools for conserving Indigenous sacred areas in Mexico.[41] Relevant collaborative agencies in the United States include BLM, EPA (in particular its Tribal Infrastructure Program), Fish and Wildlife Service, National Park Service, Forest Service, and Bureau of Indian Affairs. Collaborative agencies and organizations in Mexico include the Secretaría de Medio Ambiente y Recursos Naturales (Secretariat of Environment and Natural Resources), the Comisión Nacional para el Conocimiento y Uso de la Biodiversidad (National Commission for the Knowledge and Use of Biodiversity), the Comisión Nacional para el Desarollo de Pueblos Indígenas (National Commission for the Development of Indigenous Peoples), and Pronatura.

National Heritage Area Designations

The U.S. Congress established the Yuma Crossing National Heritage Area in 1999. This designation gives extra protection to the cultural and environmental resources found in this area. As recommended by the Good Neighbor Environmental Board, National Heritage Areas can be created along the border, including sections of Tohono O'odham land in Arizona.[42] These designations must be enforced by U.S. legislation (see chapter 15 in this volume on the REAL ID Act).

Conclusions

Indigenous nations and communities are impacted by the same environmental issues that other communities face: vegetation loss and contamination, water quality and quantity, migration impacts, industrial development impacts, and so on. In addition, Native nations and communities place a strong emphasis on continuing their culture, which is often intrinsically bound up with the environment. Protecting the environment means protecting their culture and vice versa.

Within the U.S. federal system, there is no clearly defined legal framework for Native governments to participate in forming environmental regulation and policy. Under the Mexican federal system, there is no clearly defined legal framework for expressing the sovereignty of Indigenous communities or any system for defining the limits of their jurisdiction. This lack of institutional access on both sides of the border presents major complications to protecting the interests of Indigenous communities.

Native nations and communities along the U.S.–Mexico border have benefitted from interagency cooperation to address their environmental concerns. Further cooperation and representation at all levels of policy formation and decision making will help sustain the transboundary environment and its people for generations to come.

Notes

1. J. McCormack, L. Casseen, and J. Kariyeva, "Tribal nations of northwestern Mexico and the southwestern United States," unpublished map created for Native Nations Institute, University of Arizona, 2007; see also chapter 1 in this volume and A. Kilpatrick, M. Wilken, and M. Connelly, *Indian Groups of the California–Baja California Border Region: Environmental Issues*, Project no. IT97-1 (San Diego: Southwest Consortium for Environmental Research and Policy, California Projects, 1997), available at http://www.scerp.org/projects/Kilpatrick97.pdf, accessed July 17, 2008.

2. U.S. Department of State, "Passports: First Time" (2008), available at http://www.travel.state.gov/passport/get/first/first_832.html, accessed April 15, 2008.

3. D. H. Getches, C. F. Wilkinson, and R. A. Williams, *Federal Indian Law*, 4th ed. (St. Paul, Minn.: West Group, 1998), chap. 1; F. Pommersheim, *Braid of Feathers: American Indian Law and Contemporary Tribal Life* (Berkeley and Los Angeles: University of California Press, 1995).

4. Getches, Wilkinson, and Williams, *Federal Indian Law;* Pommersheim, *Braid of Feathers.*

5. The exceptions include Public Law (PL) 83-280, where certain states have concurrent criminal jurisdiction with tribes. California is a PL 83-280 state. See also V. J. Jiménez and S. C. Song, "Concurrent tribal and state jurisdiction under Public Law 280," *American University Law Review* 47(6) (1998): 1627–707.

6. C. E. Flores-Rodríguez, *Suelo ejidal en México: Un acercamiento al origen y destino del suelo ejidal en México*, Red Cuadernos de Investigación Urbanística no. 57 (Madrid: Departamento de Urbanística y Ordenación del Territorio, 2008).

7. F. López-Bárcenas, *Legislación y derechos indígenas en México, México*, Serie Derechos Indígenas no. 3 (Juxtlahuaca, Mexico: Centro de Orientación y Asesoría a Pueblos Indígenas, Ediciones Casa Vieja/La Guillotina, 2002).

8. N. P. Alvarado-Solís, *Pápagos*, Pueblos Indígenas del México Contemporáneo series (Mexico City: Comisión Nacional para el Desarrollo de los Pueblos Indígenas, 2007).

9. López-Bárcenas, *Legislación y derechos indígenas en México.*

10. G. P. Nabhan, *The Beginning and the End of the Colorado River: Protecting the Sources, Ensuring Its Courses* (2008), available at http://www.garynabhan.com/press/gpn000021.pdf, accessed July 14, 2008.

11. J. Morrison, "Tribal lands: Emergency aid for an ailing river," *National Wildlife Magazine* 44, no. 3 (2006): 12–13.

12. M. Wilken-Robertson, "Indigenous groups of Mexico's northern border region" and "Indigenous groups of Baja California and the environment," in *Tribal Environmental Issues of the Border Region*, edited by M. Wilken-Robertson, 31–48 and 49–70, Southwest Consortium for Environmental Research and Policy, Monograph no. 9 (San Diego: San Diego State University Press, 2004).

13. Morrison, "Tribal lands."

14. L. Reeves, "Cocopah's efforts to restore lower Colorado River limitrophe," *Border 2012: U.S.–Mexico Environmental Program Regional Workgroup Newsletter, Arizona-Sonora* (Spring 2007), available at http://epa.gov/region09/indian/features/cocopah/az_sn_news4_eng.pdf, accessed February 3, 2009.

15. Ibid.; Morrison, "Tribal lands."

16. M. Hernandez, "Team works to bring local Yuma agencies together," April 16, 2008, available at the City of Yuma Web site, http://www.ci.yuma.az.us/news_8446.htm, accessed July 18, 2008.

17. "Yuma Crossing Heritage Area," *Environmental Rehabilitation* (2009), available at the Yuma Crossing Heritage Area Web site, http://www.yumaheritage.com/enviro.html, accessed April 10, 2009.

18. "Yuma Crossing Heritage Area," in *Yuma Crossing Heritage Area: Midpoint Progress Report* (2008), 10, available at http://www.yumaheritage.com/pdf/HeritageAreaMidPointReport.pdf, accessed April 10, 2009; U.S. Border Tribes and Mexican Indigenous Communities, *Tribal Accomplishments and Issues Report* (2007), prepared for the Border 2012 National Coordinators Meeting, San Antonio, Texas, May 22–24, 2007, available at http://www.epa.gov/border2012/docs/Tribal-Rp-Final-May2007-english.pdf, accessed March 17, 2008.

19. Pronatura Noroeste, A.C., "Project site: Colorado River Delta, Sonora" (2009), available at http://www.pronatura-noroeste.org/coloradoriverdelta.php, accessed February 26, 2009.

20. California Indian Basketweavers Association (CIBA), "Working to preserve our proud heritage" (2009), available at the CIBA Web site, http://www.ciba.org/ongoingWork.html, accessed February 3, 2009.

21. W. K. Kellogg Foundation, "Food systems and rural development," in *TOCA: Past Wisdom Reinvents the Future* (2006), available at http://www.wkkf.org/default.aspx?tabid=55&CID=4&ProjCID=19&ProjID=120&NID=28&LanguageID=0, accessed February 1, 2009.

22. Good Neighbor Environmental Board (GNEB), *U.S. Mexico Border Environment: Air Quality and Transportation and Cultural and Natural Resources*, Ninth Report of the GNEB to the President and Congress of the United States (Washington, D.C.: U.S. Environmental Protection Agency, 2006), 35, available at http://www.epa.gov/ocem/gneb/gneb9threport/English-GNEB-9th-Report.pdf, accessed March 12, 2008.

23. V. Parker, "CIBA joins lawsuit to stop spraying," *Roots and Shoots* 46 (Summer–Fall 2005), available at http://www.ciba.org/newsletters/RootsShootsNo46.pdf, accessed March 18, 2008.

24. J. Kalt, "Public lands traditional gathering policy finalized," *Roots and Shoots* 48 (Fall–Winter 2006), available at http://www.ciba.org/newsletters/RootsShootsNo48.pdf, accessed March 18, 2008.

25. U.S. Department of Agriculture, Forest Service, and U.S. Department of the Interior, Bureau of Land Management, "Gathering policy" (January 29, 2006), available at http://www.naepc.com/meetings/EDPM/jan09/Gathering_Policy.pdf, accessed February 3, 2009.

26. Kalt, "Public lands traditional gathering policy finalized."

27. Ibid.

28. Morrison, "Tribal lands"; GNEB, *U.S. Mexico Border Environment.*

29. International Indigenous Treaty Council, *Final Report from the Indigenous Peoples' Border Summit of the Americas II* (2007), summit held in the San Xavier District, Tohono O'odham Nation, November 7–10, 2007, available at http://www.treatycouncil.org/section_21141711211121111322111211.htm, accessed April 11, 2008.

30. Ibid.

31. F. Grissom, "Planned border wall blocks Tiguas from sacred grounds," *El Paso Times*, May 13, 2008, available at http://www.elpasotimes.com/news, accessed April 7, 2009.

32. GNEB, *U.S. Mexico Border Environment.*

33. Derechos Humanos, "The Sonoran Desert and the U.S.–Mexico border" (2006), available at http://www.derechoshumanosaz.net, accessed September 27, 2006. See also chapters 13, 14, and 15 in this volume.

34. U.S. Bureau of Land Management, *Southern Arizona Project to Mitigate the Damages as a Result of Illegal Immigration*, 2005 end-of-year report (Washington, D.C.: U.S. Bureau of Land Management, 2005), available at http://www.blm.gov/pgdata/etc/medialib/blm/az/pdfs/undoc_aliens/05_report.Par.36934.File.dat/report_complete.pdf, accessed April 10, 2009.

35. Ibid.

36. U.S. Bureau of Land Management, *Southern Arizona Project to Mitigate the Damages as a Result of Illegal Immigration*, 2006 end-of-year report (Washington, D.C.: U.S. Bureau of Land Management, 2006), 3, available at http://www.blm.gov/pgdata/etc/medialib/blm/az/pdfs/undoc_aliens/06_report.Par.39431.File.dat/06report_complete.pdf, accessed April 10, 2009.

37. GNEB, *U.S. Mexico Border Environment;* Cultural Affairs Office, Tohono O'odham Nation, 2006.

38. U.S. Bureau of Land Management, *Southern Arizona Project* (2005).

39. Ibid.; U.S. Border Tribes and Mexican Indigenous Communities, *Tribal Accomplishments and Issues Report* (2008), 17, prepared for the Border 2012

National Coordinators Meeting, Ciudad Juárez, Chihuahua, Mexico, September 3–5, 2008, available at http://www.epa.gov/border2012/docs/Tribal-Report-Sept-2008-eng.pdf, accessed April 11, 2009.

40. Executive Orders (e.g., EO 13175) and Secretarial Orders (e.g., SO 3206) set the basis for these tribal consultation policies. See M. Sanders, *Implementing the Federal Endangered Species Act in Indian Country*, Joint Occasional Papers on Native Affairs 2007-01 (Tucson: Native Nations Institute for Leadership, Management, and Policy, and Harvard Project on American Indian Economic Development, 2007), available at http://nni.arizona.edu/resources/jopna.php; Indigenous Peoples Subcommittee of the National Environmental Justice Advisory Council, *Guide on Consultation and Collaboration with Indian Tribal Governments and the Public Participation of Indigenous Groups and Tribal Members in Environmental Decision Making* (2000), available at http://www.lm.doe.gov/env_justice/pdf/ips_consultation_guide.pdf, accessed April 11, 2009.

41. Pronatura, "Iniciativa de sitios sagrados naturales" (2009), available at http://www.pronatura.org.mx/tierras_sitios_sagrados.php, accessed July 6, 2009.

42. GNEB, *U.S. Mexico Border Environment.*

Chapter 5

The Wisdom of the Sierra Madre

Aldo Leopold, the Apaches, and the Land Ethic

Richard L. Knight

In a Nutshell

- Aldo Leopold's 1936 and 1937 journeys to the Sierra Madre Occidental's Río Gavilan provided ecological insights and realizations that were important in the development of his land ethic.
- At that time, the area may have been spared ecological deterioration because of the presence or rumors of Apaches who persisted in the region after Geronimo's capture in 1886, which thwarted loggers, miners, and ranchers for several decades, and thus provided insight into pre-European settlement ecosystem conditions.
- Today, the area's natural history still offers important insights into ecological function and natural-disturbance regimes, such as fire histories.
- Despite U.S. border infrastructure projects, local and regional natural-resource interests are working cooperatively to conserve important ecological connections that span the border.

It was here that I first clearly realized that land is an organism, that all my life I had seen only sick land, whereas here [Mexico's Sierra Madre] was a biota still in perfect aboriginal health. The term "unspoiled wilderness" took on a new meaning.
—Aldo Leopold, "Foreword," typescript, July 31, 1947, Aldo Leopold Archives, University of Wisconsin Digital Collections

Today, Aldo Leopold is acknowledged as a cornerstone in the creation of the field of nature conservation in the United States and beyond. By action and word, he defined such topics as wildlife management, ecological restoration, environmental ethics, ecosystem management, conservation advocacy, and the wilderness movement. Because of his interest in human-land relationships and his uncanny ability to write as both an ecologist and a poet, his legacy inspires and shapes conservation practice into the twenty-first century.

Aldo Leopold's land ethic, perhaps his greatest legacy to conservation, required numerous ingredients to ripen and mature. I propose that one ingredient essential to its development was his time spent hunting deer in Mexico's Sierra Madre. Why was this landscape still in "aboriginal health," even though just north of the international border in Arizona and New Mexico, where Leopold worked for nearly two decades, the land and its biota were increasingly degraded?

Geronimo and a few score Apaches had surrendered to General Nelson A. Miles in 1886, thereby ending the "Apache Wars," but Apaches continued to inhabit the Mexican Sierra Madre until the late 1930s. The presence of these Apaches was one factor in retarding the advance of loggers, miners, hunters, and ranchers who were eager to exploit the region's natural resources. In the words of Grenville Goodwin, an Apache ethnographer, "Even after Geronimo surrendered in 1886 . . . the knowledge that some Apaches remained in the Sierra Madre kept most Mexicans out of the mountains for another generation."[1]

In the late 1930s, there were few places left in North America where one could witness the full complement of plants and animals native to a region, where the ecological processes still operated within their historical range of variability, and where the human culture was reminiscent of an earlier time. Leopold, through circumstance and good fortune, chose one of those places. His land ethic is part of the legacy of that place in time.

The wars between the U.S. government and the Apaches lasted from 1861 to 1886, longer than any other Indian conflict within the United States. First General George Crook and then General Miles led cavalry guided by Apache scouts into Mexico in pursuit of Apaches who either refused to remain on U.S. reservations or who periodically bolted from them to raid Mexicans and Americans. The result of the U.S. military campaigns was that the "Indian problem" on the U.S. side of the border was mostly contained. Apaches were subdued on reservations and, by and large, adapted to the order imposed upon them by the U.S. government.

Mexico, however, still offered sanctuary to Apaches who either called the Sierra Madre home or sought refuge there after fleeing U.S. reservations. This was the situation when Geronimo—with other warriors, women, and children—fled Turkey Creek on the San Carlos Reservation on May 17, 1885, and headed south into Mexico. What followed was the last spasm of the "Apache troubles," culminating in Geronimo's final surrender to General Miles on September 4, 1886, in Skeleton Canyon in the Peloncillo Mountains of Arizona.

Repeating a long history of unfair dealing with Native Americans, General Miles promptly shipped nearly all of the more than four hundred Chiricahua Apaches, including Apache scouts who had worked for the U.S. government, to a prison in Florida. Captain John G. Bourke, who had served General Crook during these Apache struggles, wrote of the American behavior, "There is no more disgraceful page in the history of our relations with the American Indians than that which conceals the treachery visited upon the Chiricahuas who remained faithful in their allegiance to our people."[2]

So when 1887 dawned in the American Southwest, the Apaches were either in prison in Florida or on U.S. reservations, and the last of the Apaches in Mexico had exited with Geronimo's surrender. Or at least that's what people believed.

In reality, an amazing truth has emerged that not all of the Apaches walked out of the Sierra Madre with Geronimo that September in 1886. Indeed, it now appears that more Apaches stayed than surrendered.

It was this continued presence of Apaches in the Sierra Madre that, by default, kept these mountains reminiscent of what an ecosystem resembled on the "eighth day of creation." And, of course, the Apache presence allowed Leopold and his companions on trips in 1936 and 1937 to see a landscape still in "aboriginal health."

> From the beginning it had surprised me how very ignorant the people of Sonora were regarding the Sierra Madre . . . mainly due to the fact that until very recently this entire part of the sierra, from the border of the United States south about 400 kilometers (250 miles), was under the undisputed control of the wild Apache Indians. It was not even now safe for a small party to cross the Sierra Madre, as dissatisfied Apaches were constantly breaking away from the San Carlos Reservation in Arizona, and no Mexican could have been induced to venture singly into that vast unknown domain of rock and forest.[3]

In September 1936, Aldo Leopold and his friend Ray Roark visited the Sierra Madre for a two-week bowhunt.[4] Clarence Lunt, a Mormon rancher from Colonia Pacheco, guided them to the headwaters of the Río Gavilan. Chasing deer with bow and arrow, seeing and hearing evidence of wolves, mountain lions, bear, deer, turkey, and other wildlife, they probably had to pinch themselves repeatedly to appreciate the wildness and land health they were witnessing. Leopold was later to write about the guacmaja (Thick-billed Parrot) as the "numenon" of the Sierra Madre.[5] In contradiction to a "phenomenon," which is ponderable and predictable, a numenon, as

Leopold described it, is something imponderable, "the significance of which is inexpressible in terms of contemporary science."[6] Leopold was clearly moved by his experience. The following year, just before Christmas, he again returned to the Río Gavilan watershed with his brother Carl and oldest son, Starker.[7] This trip was equally memorable. The opening lines of his *Conservationist in Mexico* begin, "Our southwestern mountains are now badly gutted by erosion, whereas the Sierra Madre range across the line still retains the virgin stability of its soils and all the natural beauty that goes with that enviable condition."[8]

The sharp contrast on either side of the international boundary was not lost on Leopold: "it is ironical that Chihuahua, with a history and a terrain so strikingly similar to southern New Mexico and Arizona should present so lovely a picture of ecological health, whereas our own states, plastered as they are with National Forests, National Parks and all the other trappings of conservation, are so badly damaged that only tourists and others ecologically color-blind can look upon them without a feeling of sadness and regret."[9]

For Leopold, these two trips must have been an antidote to his 1935 trip to Germany, where he had seen the polar opposite in terms of forests and land health. In Germany, he saw endless acres covered in forest monocultures and noted, "There is an almost uncanny mixture of the admirable with the false in everything one sees here."[10] Leopold wrote about this visit, "We Americans have not yet experienced a bearless, wolfless, eagleless, catless woods. We yearn for more deer and more pines, and we shall probably get them. But do we realize that to get them, as the Germans have, at the expense of their wild environment and their wild enemies, is to get very little indeed?"[11]

Leopold's two visits to the Río Gavilan watershed, coupled with his other experiences, were clearly pivotal to his land consciousness. He seemed to understand that by visiting the Sierra Madre, he might find a watershed still in its proper functioning condition, but he also seemed to appreciate that this opportunity was lost on the U.S. side of the border due to the heavy hand of the "economic juggernaut." Because of a differing economy, history of governance, and culture, Mexico was able to offer Leopold something he could not find in his own country, the nation that gave the world not only its first national park and wilderness, but also a conservation movement!

Starker Leopold, Aldo and Estella's oldest son, revisited the Río Gavilan watershed in 1948, and the Apaches were still on his mind, for he wrote, "I even wondered if a few of Geronimo's Apaches might remain in some hidden barranca—a thought that made my scalp tingle one early morning when a mysterious column of smoke rose from the canyon below me and then was snuffed out as suddenly as it had appeared."[12]

Alas, neither the Apaches nor "aboriginal health" was left in the Río Gavilan, for by the time of Starker's visit the heavy hands of loggers and hunters had laid hold of the region. Starker wrote, "At once it was evident that a great change had come to the Sierra Madre. The narrow, rocky road of eleven years back was graded and ditched, and around almost every bend we met a lumber truck groaning down the grade under a staggering load of pine planks."

In this heart-rending essay of loss, "Adios, Gavilan," Starker acknowledged his father's greatest fear and the seemingly inevitable march of "progress": "the loggers had not reached the particular area where we hoped to camp, but they were working in that direction from three sides. We knew then that instead of initiating an era of renewed acquaintance with the wilderness, we had come to witness its passing."

In saying good-bye to his youthful dreams of fulfilling his father's dream, Starker ended "Adios, Gavilan" with these words: "I would like to think that there is another river filtering off hillsides of golden grama and winding under virgin pine. May my son some day explore its rimrocks and imagine there are still Apaches in its tributaries."[13]

How are the Sierra Madres today, in the first decade of the twenty-first century, and is this the last chapter in the story of Aldo Leopold, land health, the Sierra Madre, and life straddling the international "Tecate Line"?

Today, Aldo and Estella Leopold's great-grandson Jed Meunier has returned to these same mountain ranges, and like the Leopolds before him, he comes to study the intricacies of human-land relationships on either side of the international boundary. Jed, a graduate student at Colorado State University, is examining how decades of fire suppression have disrupted the historical range of variability in ecosystem processes, including fire. There is still the opportunity on the Mexican side of the line to discover Leopold's "base datum of normality" in fire histories. An example of Madrean forest health cannot be found on the national forests north of the U.S.–Mexico border due to fire suppression, a by-product of wealth and technology in the United States. Indeed, decades of combating wildfires has so altered the historic fire patterns that our national forests would, as in Leopold's day, best be categorized as "sick."

Interestingly, Jed's ability to conduct his studies is greatly facilitated by two nongovernmental organizations that place a premium on land health and international cooperation. The Animas Foundation owns the Animas Mountain Range in New Mexico, and the Cuenca Los Ojos Foundation controls large holdings in the Sierra San Luis, across the line in Mexico. Both foundations believe in natural fire and do what they can to allow fires, once started, to burn as topography and weather wish. Their good work has been supported by the million-acre "working wilderness" landscape of the Malpai Borderlands Group.[14]

These groups are collectively interested in building bridges across the dry borderlands, in stark contrast to the U.S. policy of building walls along 1,100 kilometers (700 miles) of the U.S. border with Mexico. Whereas the Animas Foundation, the Malpai Borderlands Group, and Cuenca Los Ojos Foundation are interested in keeping the border porous for dispersing jaguars and vital ecological processes, such as fire, the U.S. government sees the border as a threat and hopes that a wall will keep people and everything else out.

Border walls seem to violate a deep sense of identity that most Americans cherish. As a nation of immigrants with our own goddess of welcome, the Statue of Liberty, we all are left with divided feelings about what a wall says about us as a nation. Perhaps, as with so many things of great import, we can permit Aldo Leopold to have the last word: "It is a fact, patent both to my dog and myself, that at daybreak I am the sole owner of all the acres I can walk over. It is not only boundaries that disappear, but also the thought of being bounded."[15]

Notes

1. N. Goodwin and N. Goodwin, *Apache Dairies: A Father-Son Journey* (Lincoln: University of Nebraska Press, 2000), 5–6.

2. J. B. Bourke, *On the Border with Crook* (Lincoln: University of Nebraska Press, 1971), 485.

3. C. Lumholtz, *Unknown Mexico: A Record of Five Years' Exploration among the Tribes of the Western Sierra Madre; in the Tierra Caliente of Tepic and Jalisco; and among the Tarascos of Michoacán* (New York: Charles Scribner's Sons, 1902), 23–24.

4. C. Meine, *Aldo Leopold: His Life and Work* (Madison: University of Wisconsin Press, 1988).

5. A. Leopold, *A Sand County Almanac and Sketches Here and There* (New York: Oxford University Press, 1949), 38.

6. Meine, *Aldo Leopold*, 368.

7. Ibid.

8. A. Leopold, "Conservationist in Mexico," *American Forests* 43 (1937), 118.

9. Ibid.

10. Meine, *Aldo Leopold*, 356.

11. A. Leopold, "Naturschutz in Germany," *Bird-Lore* 3(2) (1936), 110.

12. A. S. Leopold, "Adios, Gavilan," *Pacific Discovery* 2(1) (1949), 5.

13. Ibid., 5, 13.

14. N. F. Sayre, *Working Wilderness: The Malpai Borderlands Group and the Future of the Western Range* (Tucson, Ariz.: Rio Nuevo Publishers, 2005).

15. A. Leopold, *A Sand County Almanac*, 41.

Conservation of Shared Species

Conservation of Shared Species

The United States and Mexico have in common many species of animals that move regularly across their border. Birds, bats, and insects migrate every year from Central America and Mexico to the United States and Canada, and then back again. In addition, species such as black bear, flat-tailed horned lizards, Mexican wolves, and Sonoran pronghorn regularly range across the border. In order to move throughout their ranges, whether traveling by land or flying through the skies, wildlife need connected habitats that span the border. Animals moving overland need safe corridors through which to travel, and winged creatures need stopover areas in which to rest and refuel during their flights. To ensure habitat connectivity for the species that move across the border, binational, coordinated efforts are needed.

In a 1999 book entitled *Continental Conservation*, conservation biologists Michael Soule and John Terborgh suggest that a continentwide network of reserves should be designed to connect species habitats throughout North America. Although they thoughtfully lay out the biological necessity for continentwide efforts to promote habitat connectivity, they are less detailed about how to effect these efforts.[1] Protecting habitat for species with continental distributions requires working across international borders and entails all the administrative complexities of simultaneously working in different countries. The goal of the transboundary conservation efforts described in this section's chapters is to preserve the connectivity of habitats for species whose natural-range distributions cross the U.S.–Mexico border. In addition, the authors offer concrete suggestions for how to conduct collaborative conservation across borders.

The three chapters in this section address the issue of transboundary wildlife movement at increasing geographic scales. Melanie Culver and her colleagues discuss species that reside in the U.S.–Mexico border region. Charles Chester and Emily McGovern consider species with broader continental distributions, and José Bernal Stoopen and his coauthors examine species with both broad distributions and narrow borderland ranges.

The section begins with chapter 6 and its explanation of why habitat connectivity is important for maintaining genetically healthy wildlife populations. Using several borderland species as examples, the authors discuss how habitat fragmentation and barriers such as the border wall inhibit wildlife movement, in turn impeding gene flow and threatening species survival. They detail how drivers such as grazing, land-use conversion, and climate change are fragmenting wildlife habitat in the borderlands. They argue that binational coordination is needed to identify and conserve cross-border corridors for wildlife movement.

Chester and McGovern consider in chapter 7 the challenges of protecting birds that migrate between Mexico and the United States. They point out that protecting birds with expansive, continental distributions is extremely complex because it requires protecting summer breeding ranges, wintering grounds, and migratory stopover sites in between. They review the twenty-plus U.S.–Mexico conservation initiatives aimed at protecting migratory birds and conclude by suggesting that although integration across all these diverse initiatives is needed, it is also critical to protect more habitat.

And, finally, chapter 8 examines the organizational and social conditions necessary for successful binational conservation of transboundary endangered species. The authors surveyed participants in binational recovery programs for endangered species, focusing mainly on efforts to protect the Mexican wolf. They determined that successful binational recovery programs require sustained funding, multilevel government agency coordination, equitable participation between Mexican and U.S. stakeholders, and personnel continuity.

Note

1. M. E. Soule and J. Terborgh, *Continental Conservation: Scientific Foundations of Regional Reserve Networks* (Washington, D.C.: Island Press, 1999).

Chapter 6

Connecting Wildlife Habitats across th
U.S.–Mexico Border

Melanie Culver, Cora Varas, Patricia Moody Harveson,
Bonnie McKinney, and Louis A. Harveson

In a Nutshell

- Genetic diversity in wildlife populations is critical for long-term survival and is maintained by species movement between connected habitats through migration corridors.

- Human-induced habitat fragmentation impedes wildlife migration, which isolates wildlife populations and increases the risk of extinction.

- Habitat fragmentation can be caused by ranching practices, land-ownership practices, predator-control practices, land-use conversion, and climate change.

- Species survival depends on connected cross-border habitats, as demonstrated in case studies of black bears, mountain lions, jaguars, ocelots, bobcats, and flat-tailed horned lizards.

- The case studies indicate that species cross political boundaries, so binational coordination at local and national levels is required to ensure that transborder wildlife movement corridors are identified and conserved to maintain habitat connectivity.

Introduction

A major challenge in U.S.–Mexico transboundary wildlife conservation is maintaining habitat connectivity for wildlife movement across the international border. Wildlife habitat in the border region is being transformed and fragmented by human land-use and management practices. These habitat changes create barriers to wildlife movement across the border, and the situation is further complicated by differences in land-use and management practices between the United States and Mexico.

In this chapter, we review the drivers of habitat change and border region fragmentation to explain how they create barriers to wildlife movement. Using six case studies, we explain how habitat fragmentation and barriers impede species movement and gene flow, and why maintaining habitat connectivity and gene flow is essential to species survival. We argue that maintaining gene flow is critical when wildlife habitat boundaries do not coincide with political boundaries. Finally, we reflect on the prospects for collaborative conservation of wildlife and detail the challenges for reversing habitat fragmentation.

This chapter focuses on the border region shared by Arizona, New Mexico, and Texas in the United States with Sonora, Chihuahua, and Coahuila in Mexico. The region contains the Madrean Archipelago, an intact ecoregion spanning the Sonoran and Chihuahuan deserts, containing more than twenty-seven mountain ranges surrounded by desert, and influenced by tropical and temperate climates. The Madrean Archipelago is extraordinarily diverse ecologically and is home to many endemic species. Most of the U.S. Southwest's game, threatened, and endangered species are present in this region.

Background

The Importance of Maintaining Genetic Diversity and Habitat Connectivity

According to the International Union for Conservation of Nature and Natural Resources, conserving genetic diversity is one of three global conservation priorities. In the U.S.–Mexico border region, a primary wildlife-management goal is to maintain high population-level genetic diversity. Populations with high genetic diversity can adapt to changing environments, but small populations fragmented by land-use changes are particularly vulnerable to being cut off from immigration and emigration and to losing genetic diversity over time (fig. 6.1). These populations can have lower reproductive fitness and greater risk of extinction due to inbreeding.[1]

Human-induced habitat fragmentation may divide a population into spatially separated subpopulations.[2] The survival of a subpopulation depends on the size of the patch, the amount of edge, and the suitability of corridors for movement among patches.[3] A suitable corridor must be large enough and with a habitat quality high enough so that connectivity is created among habitat patches to allow for animal movement through

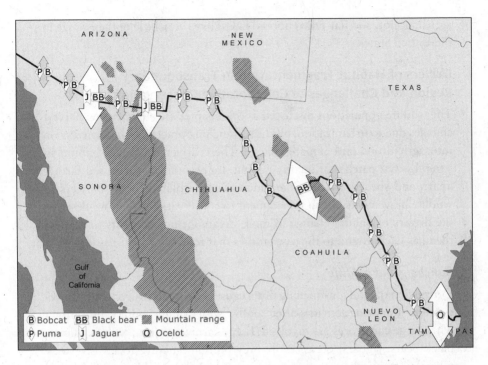

Figure 6.1. Map of the transboundary region including mountain ranges and political borders. Arrows indicate areas of importance for connectivity of black bear (BB), puma (P), jaguar (J), bobcat (B), and ocelot (O). These areas are important for binational habitat protection. Map drawn by Mickey Reed and Renee LaRoi.

the corridor. For example, barriers such as highways are rarely crossed by small rodents but are occasionally crossed by large cats or bears.[4] However, larger barriers along parts of the U.S.–Mexico border—such as walls, infrastructure (e.g., border easement roads, barricades, staging areas, lights), and increased human activity (e.g., law enforcement and maintenance operations)—have the potential to stop many terrestrial animals' movement. Border easement roads are up to 18 meters (60 feet) wide in some places along the wall.

Maintaining connectivity among subpopulations is particularly important for species with a small portion of their range on one side of the international border because this small subpopulation is vulnerable to genetic isolation and loss of diversity over time. Species with a very small part of their range in Mexico are bobcat and black bear. In contrast, the jaguar, ocelot, and flat-tailed horned lizard have very little of their range in the United States.

Drivers of Habitat Fragmentation in Transboundary Region and Challenges to Collaborative Conservation

The wildlife populations discussed in this chapter are experiencing habitat changes due to urbanization, modified ranching practices, land conversion into agriculture, and climate change. These drivers fragment habitat in several ways: patches of quality habitat are becoming smaller and farther apart, and species are becoming increasingly isolated. These barriers to wildlife movements are an impediment to conservation. Next, we describe the drivers of habitat change in more detail, including efforts to address them, as background to the case studies that follow.

Ranching and wildlife

Conversion of land for ranching often causes habitat destruction and deforestation.[5] Ranching activities often conflict with wildlife-conservation efforts when large carnivores are involved. Large carnivores (e.g., wolves, pumas, black bears, jaguars) occasionally prey on domestic livestock.[6] Livestock mortality from other causes usually exceeds that caused by predators, but predators are often blamed. Programs to ameliorate conflicts between ranching and carnivores have recently fostered appreciation of carnivores among ranchers (e.g., the Cheetah Conservation Fund in Namibia used education and incentives for ranchers to stop killing cheetahs; in the southwestern United States, Defenders of Wildlife has a photograph incentive program to provide monetary rewards for jaguar photos on ranches).[7]

Voluntary ranching practices that benefit wildlife also exist. One example, the Malpai Borderlands Group, consists of landowners, stakeholders, and scientists who work to restore lands along the Arizona-Sonora border to benefit both ranching and wildlife.[8] Another is the collaboration between CEMEX, a Mexican cement company, and the Texas Parks and Wildlife Department to restore wildlife and habitats in Maderas del Carmen mountain range in Mexico, and on private lands adjoining Big Bend National Park and Black Gap Wildlife Management Area in Texas.[9]

Land tenure

Land ownership differs between Mexico and the United States. Mexico has very little public land; most rural land belongs to ranches, farms, or communal lands called *ejidos*. Most lands along the border in Coahuila,

Mexico, historically contained large cattle and horse ranches. Post–Mexican Revolution land reform partitioned large ranches into smaller ejidos, and Mexican citizens from other areas moved to the ejidos. This reform fragmented the landscape and caused high rates of subsistence hunting, overgrazing, and exploitation. Recent policy changes now permit ejido land to be sold, allowing new conservation initiatives by governmental agencies, nongovernmental organizations, corporations, and individuals to buy and restore wildlife habitats in northwest Mexico.[10]

Despite the growing threats of urbanization and the associated habitat fragmentation and loss, approximately 76 percent of the Chihuahuan Desert in Texas is still classified as farm or ranch land. However, land-ownership patterns in Texas are changing,[11] with many traditional ranches being sold for recreational purposes. Although these changes may fragment large tracts of land, the new landowners tend to support wildlife management and habitat restoration. Similar data are not easily accessible for the Mexican portion of the Chihuahuan Desert.[12]

Predator control

Predator control has been practiced in North America since the arrival of Europeans, who perceived threats from large carnivores to human life and livelihood.[13] It resulted in the extirpation of mountain lions from the eastern half of the United States (except for Florida) and of wolves and brown bears from the lower forty-eight U.S. states (except for gray wolves on Michigan's Isle Royale National Park and brown bears in Montana and Wyoming). In the United States, the passage of the federal Animal Damage Control Act in 1931 authorized the eradication of predators and other nuisance animals, including through paid bounties. In 1973, however, the passage of the Endangered Species Act (ESA) signified a shift in attitudes toward carnivore conservation.[14] Despite this shift, in the attempt to protect the livestock industry, heavy predator control has depleted or eliminated most large carnivores in Texas.[15] Even today, carnivore-conservation programs are controversial. The reintroduction of gray wolves into Montana, Wyoming, and Idaho (including Yellowstone National Park) and of Mexican wolves into Arizona and New Mexico, as well as the protection of jaguars in the U.S.–Mexico borderlands, are met with heavy opposition by local ranchers.

Similar predator-control activities occurred in Mexico, including paid bounties, which are still common for the removal of mountain lions,

particularly in the northern, cattle-ranching states. However, small remnant populations of black bear and mountain lion remain in fragmented
habitat in northern Mexico. In fact, the remaining black bear and mountain
lion populations in western Texas are subpopulations of northern Mexico
populations.[16]

Case Studies: Status of Transboundary Species

The species discussed in this section are a small sampling of the many transboundary species in need of conservation efforts. All occur in Mexico and
the United States. All, except the mountain lion, are rare. We have chosen
species with ranges asymmetrically distributed across the border to illustrate
the need for binational conservation efforts. A small population at the edge
of its species' range in one country (such as jaguars in the United States and
black bears in Mexico) will require frequent infusion of individuals from
the larger "source" population in the other country. For this infusion to
occur, habitat connectivity across the border is needed, requiring binational
conservation cooperation.

Black Bear

Black bears (*Ursus americanus*, see fig. 6.2) were historically distributed
widely throughout North America, in all forested habitats. They are pres-
ently found in northern Mexico, thirty-two states in the United States, and
in all Canadian provinces except Prince Edward Island. Black bears are
endangered in Mexico, the southern terminus of their range.

By the early 1900s, black bears had been overexploited throughout much
of their U.S. range. Their populations began to recover following the establishment of state wildlife agencies, wildlife regulations, and federal lands.[17]
Nonetheless, in Texas, with few harvest limits and little public land, the
black bear was considered extirpated by the 1950s, with only an occasional
sighting in lower Big Bend. Factors causing the dramatic decline include
unregulated hunting, predator control, and habitat loss. In 1987, the black
bear was listed as endangered in Texas and is currently "state threatened"
in the western half of the state.[18] Although some populations in Arizona
were nearly extirpated, hunting permits are currently issued for black bear
in Arizona and New Mexico.[19]

Reductions to the black bears' range have also occurred in Mexico,
where historically the black bear was distributed in nine Mexican states
but is currently found in only eight states: Coahuila, Chihuahua, Sonora,

Figure 6.2. Black bears. Photograph by Jonas Delgadillo.

Tamaulipas, Durango, Zacatecas, Nuevo León, and San Luis Potosí. Maderas del Carmen and adjacent Mexican mountain ranges support the largest populations of black bears.[20] Aldo Starker Leopold (Aldo Leopold's son) attributed the decline of black bears in Mexico to uncontrolled hunting, indiscriminate killing, and loss of habitat.[21] In 1986, Mexico officially listed the black bear as endangered.

Black bear habitat is typically coniferous and broadleaf deciduous woodlands, but the bears also use midelevation yucca-sotol grasslands in the borderlands. In the mountains or "sky islands" of the borderlands, mountainous bear habitat is separated by a "sea" of desert habitat. Bears must cross the lower desert habitat to dispersal between sky island mountaintops. Most sky islands are too small for a self-sustaining black bear population, with the exception of Maderas del Carmen and adjacent mountain ranges in northern Coahuila, Mexico, which act as a mainland for dispersal to the smaller sky islands.[22] An impermeable border wall would eliminate this transboundary movement vital to the survival of sky island black bears. Contiguous tracts of land with protection of dispersal corridors in western

Texas–northern Coahuila and Arizona-Sonora are imperative to sustain these black bear populations.

Due to conservation efforts and habitat inaccessibility to people, black bear populations are recovering in the isolated mountains of northern Coahuila and Texas. The population has expanded through natural dispersal into western Texas and adjacent areas in Mexico, where populations were extirpated.[23] Mexican black bears, being at the southern end of the species' range, are dependent on connectivity to U.S. populations, and, conversely, Texas is dependent on the Mexican bear population. Using mitochondrial DNA genetic studies, gene flow has been shown to occur between sky islands on either side of the U.S.–Mexico border. David Onorato and Eric Hellgren pinpointed northern Mexico as the source of black bear populations in western Texas, and Cora Varas and her colleagues showed that black bears historically dispersed from Mexico into Arizona.[24] Onorato and Hellgren suggest that four factors will lead to black bear recolonization in Texas: increased human tolerance, enforcement of hunting restrictions, availability of suitable habitat, and the presence of a mainland population.[25] Both emigration and immigration play a large role in dispersal and normal population expansion.

Mountain Lion

Mountain lions (*Puma concolor*, see fig. 6.3), also known as pumas or cougars, historically ranged throughout North and South America. However, populations throughout eastern North America were severely reduced by the late 1800s due to human actions, and western populations were diminished by the early 1900s.[26]

Until 1965, management of mountain lions was focused on eradication, not on conservation. From 1965 to 1972, western states began to regulate the harvest of mountain lions; the exception is Texas, where harvest is unlimited due to the lion's status as "vermin."[27] This regulation came on the heels of changing public attitudes toward the environment and the enacting of a variety of environmental legislation, including the ESA and the ban on 1080 (a poison used to eradicate predators). Today, mountain lions are protected in the western United States (except Texas) as a game animal, with limited harvest based on quotas or permits, yet overharvest is still a threat to mountain lion populations. Overharvesting increases extinction risks for small populations and can destabilize regional populations.[28] Overexploited mountain lion populations often exhibit altered age and sex structures compared to naturally regulated populations.

Figure 6.3. Mountain lion. Photograph by Steve Howe.

Harvest and predator control are significant causes of mountain lion mortality in the Chihuahuan Desert of Texas and Mexico. In Texas, mountain lions are nongame animals with no harvest restriction.[29] Humans are the major cause of mortality for mountain lions in the Big Bend ecoregion of Texas. In a recent study on Big Bend Ranch State Park, fifteen of sixteen radio-collared mountain lions were killed on adjacent private lands as a result of predator control.[30] Existing refugia (state or federal lands without harvest) appeared to have little effect on the survivability of mountain lions, indicating that larger refugia, or source populations, are needed for continued persistence. In Mexico, traps are prohibited, and mountain lions cannot be hunted without a permit from the Secretaría de Medio Ambiente y Recursos Naturales (Secretariat of Environment and Natural Resources), yet they are illegally killed in remote areas to curb depredation on deer and domestic livestock. The lack of protection in Texas and of law enforcement in Mexico poses risks to the continued persistence of mountain lions in the Chihuahuan Desert.

Mountain lions occur at low densities due to their large, nonoverlapping home ranges and thus depend on connectivity to neighboring populations to maintain populations large enough to avoid inbreeding.[31] Large dispersal distances facilitate connectivity to neighboring populations, but recent human development has led to loss of corridor areas.[32] Kenneth Logan and Linda Sweanor identify habitat loss as a major threat to mountain lions and suggest that habitat conservation should include protecting habitat patches and connectivity among them.[33] This connectivity is especially important where source populations contribute individuals to smaller subpopulations.

Mountain lion populations in Texas, Arizona, and Mexico are highly dependent on connectivity to neighboring populations. In Texas, mountain lions occur in two geographically separate regions (western and southern). Molecular genetic techniques have been widely used to examine relatedness among wildlife populations. With mitochondrial DNA techniques, Christopher Walker and colleagues found that the relatedness of lions in southern and western Texas was minimal and suggested that Mexico was the likely source of western Texas populations.[34] A Texas study also documented lion home ranges as large as 2,414 square kilometers (1,500 square miles), some of the largest ever recorded, further emphasizing the necessity to manage large tracts of land for this species.[35] In southern Arizona, lions have moved historically among the "sky island" mountain ranges using riparian and desert valley corridors. Brad McRae and colleagues found evidence of gene flow among mountains of the "sky island" region, indicating connectivity in this region.[36] Loss of corridors in the sky island region due to human development can isolate populations from needed gene flow through dispersal. If isolated populations are small, inbreeding and associated inbreeding depression can result—as seen in the Florida population of the Florida panther *(Puma concolor couguar)*.[37] Loss of connectivity among populations, including between Mexico and the United States, has the potential to affect mountain lion conservation negatively.

Jaguar

Jaguar *(Panthera onca)* populations have been reduced or extirpated from most of their historic distribution, which extended as far north as the Grand Canyon.[38] They currently occur from southern Arizona to Paraguay and Argentina, with most populations reduced in size due to habitat loss. In Mexico, jaguars have a discontinuous distribution and occupy approximately 65 percent of their historical range. They were considered extirpated in the United States until two hunters photographed jaguars on two different occasions in 1996. These pictures were taken several hundred miles apart in Arizona. An extant population of jaguars occurs in Sonora, Mexico, at the confluence of the Bavispe and Aros rivers, approximately 225 kilometers (140 miles) south of Douglas, Arizona.[39] All jaguar sightings in Arizona over the past ten years have been males, so these jaguars could be dispersing individuals from the Bavispe-Aros Mexico population.[40] Jaguars exhibit large home-range sizes and long-distance dispersal. Gene flow is facilitated through young males' dispersal long distances from their natal areas, so jaguars require large areas of connected habitat to maintain sustainable populations. The existence of jaguars in Arizona is entirely dependent on maintaining connectivity with the Sonora population, so an impermeable border wall will negatively affect or eliminate Arizona jaguars.

Cattle ranching remains an important livelihood in the U.S.–Mexico transboundary region. Jaguars are often removed from an area when livestock losses happen, yet it is unknown what percentage of jaguars can be removed from the wild without affecting long-term survival. More research is needed to understand fully the causes of the jaguar-livestock conflict, to increase jaguar survival in these areas, and to establish better livestock husbandry practices. Jaguars (and mountain lions) are considered keystone species, so their presence is an indicator of ecosystem health: in other words, their removal from the ecosystem would be expected to cause linked extinctions of other species.[41]

Ocelot

The ocelot *(Leopardus pardalis)* is endangered throughout its range from southern Texas to northern Argentina. The U.S. population contains fewer than one hundred ocelots in two populations in southern Texas—the northern limit of the ocelot's distribution.[42] The Texas populations were historically connected with neighboring populations in Mexico, but human activities (farming and ranching) have caused the populations to shrink and become more fragmented. Recent genetic studies show no evidence of recent gene flow between Mexico and the United States. The Texas population is sufficiently small that the U.S. Fish and Wildlife Service recovery plan for ocelots (written to identify how to alleviate threats and to promote recovery) specifies that reestablishing historical connectivity between the populations in the United States and Mexico is essential for the ocelot's long-term survival in Texas. In cases where natural connectivity is not possible, artificial connectivity is employed through translocations (physically transporting individuals from one population to another). Ocelots have already been transported from Tamaulipas, Mexico, to Texas. Reestablishing ocelot connectivity between Mexico and the United States would help ensure this species' long-term survival in Texas.

Ocelots historically occurred in Arizona, although they have been extirpated from that state.[43] The last documented sighting was in the Huachuca Mountains of Arizona in the 1960s. They were considered extirpated from Sonora in Mexico but have recently been documented in the Bavispe-Aros region of Sonora and even more recently close to the U.S.–Mexico border near Cananea by the Sky Island Alliance.[44]

Bobcat

Bobcats occur in North America from southern Canada to Mexico. Only the "Mexican bobcat" subspecies *(Lynx rufus escuinapae)* of central Mexico (the southern terminus of its range) is listed as endangered, in part due to the paucity of knowledge regarding its status and biology.[45] In 2005, based on a petition from the National Trappers Association, the Mexican bobcat was delisted under the ESA. In 1982, all bobcats were listed in the Convention on International Trade in Endangered Species due to the inability to distinguish pelts of *L. r. escuinapae* and pelts of other subspecies. The bobcat plays an integral role in the richness of biodiversity in biomes where they occur. Bobcats are underrepresented in the scientific literature in the southwestern United States and in Mexico, so there is a great need for more scientific research on this species. They are expected to become increasingly threatened as human populations continue to grow and development intensifies.[46] Like black bears, bobcats occur throughout North America, with the southern end of their range in Mexico, so they are likely to fare poorly in Mexico if a nonpermeable wall isolates them from dispersal opportunities with U.S. populations.

Flat-tailed Horned Lizard

The flat-tailed horned lizard *(Phrynosoma mcallii)* occurs in southwestern Arizona, southeastern California, and northern Mexico. It is endangered in Mexico and proposed as threatened in the United States. Its limited distribution is divided by the Colorado River, so the species is already subjected to natural fragmentation. Populations on either side of the river are fairly robust. The current Arizona population includes only a small area near Yuma, Arizona, which is separated from the California population by the Colorado River. Therefore, its only source of new individuals is from Mexico. Genetic analyses have indicated detectable levels of gene flow between the Arizona and Mexican populations.[47] Additional fragmentation might present a problem for the population in Arizona, particularly if it disrupts current gene flow from Mexican populations to Arizona populations. A disruption of this gene flow by a nonpermeable border wall might confine the Arizona population to a small, isolated habitat patch with no opportunity for continued gene exchange with populations in Mexico.

Future of Binational Habitat Protection

It is by now a maxim that transboundary conservation requires a binational perspective. The movements of organisms, as demonstrated in this chapter, do not recognize borders. Biodiversity conservation requires a landscape approach irrespective of political boundaries.[48] Research priorities should arise from a proposed binational U.S.–Mexico consortium (in addition to the trilateral North American Free Trade Agreement among the United States, Mexico, and Canada). Regional working groups can be formed to focus on small conservation efforts; however, communication between groups will be necessary for coordination. Most current research studies focus on either the U.S. or Mexico side of the border; however, truly international ecological studies might be achieved using sister studies (e.g., two coordinated and concurrent studies). Two obvious research priorities to be developed through the proposed consortium are transborder movement and corridor use. Once corridors are identified, they should be prioritized for conservation, including mitigation of identified negative effects of the border wall.[49] Mitigation may include purchase of land, restriction of public access to the border easement road, conservation easements, habitat improvement, and funded research.

Species historically formerly moved freely across international borders, but now human-related developments are blocking natural corridors in the U.S.–Mexico transboundary region. To maintain connectivity of wildlife populations through newly created barriers, a multidisciplinary effort including biology, ecology, policy, laws, and human behaviors and perceptions is critical. In transboundary conservation, there is the added requirement of international cooperation and involvement of all countries that share borders relevant to a specific conservation priority. Joint participation fosters consistent conservation actions on both sides of the border. These international conservation efforts and the conservation biologists involved will in the process strive to address continually changing conservation needs as populations, habitats, environmental factors, and human behaviors change.

Transboundary wildlife conservation efforts also need to occur on several scales. It is essential to work at a local scale within the habitats where species live, but conservation, to be effective, must also take place at state and federal governmental levels. In other words, conservation cannot succeed if local people are not supportive; however, governmental agency collaboration is also critical. In the U.S.–Mexico transboundary region, assistance from the U.S. Department of Homeland Security and U.S. Border Patrol can be a key element of success. Cooperation among Mexico, the United States, and Canada must occur to ensure connectivity among subpopulations and across countries.

The examples presented in this chapter are species that are abundant on one side of the border and less common on the other. This imbalance puts the less-abundant population at risk of extirpation if connectivity is detached across the border. Of the six species discussed in this chapter, the mountain lion is the exception because it is common on both sides of the border, and its range includes all of Mexico and farther south. For that reason, the mountain lion is a perfect model to examine crossing points and the extent of cross-border gene flow and thus to design permeability into the border wall (if building of the wall continues). Precise movement data on mountain lions might result in recommendations for modifying the border wall and easement roads to improve habitat quality and connectivity, supporting wildlife populations that span the border.

Acknowledgments

The authors thank Ron Thompson for his input on all aspects of this chapter, as well as Raul Valdez, Lisa Haynes, Tony Dee, Karla Pelz Serrano, Samia Carrillo-Percastegui, Judith Ramirez, Terry Myers, and Hans-Werner Herrmann for their helpful comments on the chapter's content.

Notes

1. R. Frankham, J. D. Ballou, and D. A. Briscoe, *Introduction to Conservation Genetics* (Cambridge, U.K.: Cambridge University Press, 2002).

2. G. K. Meffe and C. R. Carroll, *Principles of Conservation Biology* (Sunderland, Mass.: Sinauer Associates, 1997).

3. P. Beier, "Determining minimum habitat areas and corridors for cougars," *Conservation Biology* 7 (1993): 94–108.

4. Meffe and Carroll, *Principles of Conservation.*

5. T. Sheridan, "Cows, condos, and the contested commons: The political ecology of ranching on the Arizona-Sonora borderlands," *Human Organization* 60 (2001): 141–52.

6. O. C. Rosas-Rosas, L. C. Bender, and R. Valdez, "Jaguar and puma predation on cattle calves in northeastern Sonora, Mexico," *Rangeland Ecology Management* 61 (2008): 554–60.

7. L. Marker, Cheetah Conservation Fund, personal communication, September 2007; Craig Miller, Defenders of Wildlife, personal communication, November 2006.

8. B. McDonald, "A private landowner's perspective: Conservation biology and the rural landowner," in Meffe and Carroll, *Principles of Conservation Biology*, 21.

9. B. R. McKinney and J. A. Delgadillo, "Preliminary report on Maderas del Carmen black bear study, Coahuila, Mexico," *Western Black Bear Workshop* 9 (2007): 21–28.

10. R. Valdéz, J. C. Guzmán-Aranda, F. J. Abarca, L. A. Tarango-Arambula, and F. C. Sánchez, "Wildlife conservation and management in Mexico," *Wildlife Society Bulletin* 34 (2006): 270–82.

11. N. Wilkins, A. Hays, D. Kubenka, D. Steinbach, W. Grant, E. Gonzalez, M. Kjelland, and J. Shackelford, *Texas Rural Land: Trends and Conservation Implications for the 21st Century*, Publication B-6134, Texas Cooperative Extension (College Station: Texas A&M University, 2003).

12. A. S. Leopold, *Wildlife of México: The Game Birds and Mammals* (Berkeley and Los Angeles: University of California Press, 1959).

13. J. K. Yoder, "Contracting over common property: Cost-share contracts for predator control," *Journal of Agricultural and Resource Economics* 25 (2000): 485–500.

14. M. J. Robinson, *Predatory Bureaucracy: The Extermination of Wolves and the Transformation of the American West* (Boulder: University Press of Colorado, 2005).

15. D. J. Schmidly, *Natural History of Texas: A Century of Change* (Lubbock: Texas Tech University Press, 2003).

16. C. W. Walker, L. A. Harveson, M. T. Pittman, M. E. Tewes, and R. L. Honeycutt, "Microsatellite variation in two populations of mountain lions *(Puma concolor)* in Texas," *Southwestern Naturalist* 45 (2000): 196–203; D. P. Onorato, E. C. Hellgren, R. A. Van Den Bussche, and D. L. Doan-Crider, "Phylogeographic patterns within a metapopulation of black bears *(Ursus americanus)* in the American Southwest," *Journal of Mammalogy* 85 (2004): 140–47.

17. M. R. Pelton, "Black bear," in *Ecology and Management of Large Mammals in North America*, edited by S. Demaris and P. R. Krausman, 389–408 (Upper Saddle River, N.J.: Prentice Hall, 2000).

18. D. P. Onorato and E. C. Hellgren, "Black bear at the border: Natural recolonization of the Trans-Pecos," in *Large Mammal Restoration: Ecological and Sociological Challenges in the 21st Century*, edited by D. S. Maehr, R. F. Noss, and J. L. Larkin, 245–59 (Washington, D.C.: Island Press, 2001).

19. A. L. LeCount and J. C. Yarchin, *Black Bear Habitat Use in East Central Arizona*, Technical Report no. 4 (Phoenix: Arizona Game and Fish Department, 1990).

20. B. R. McKinney and J. A. Delgadillo Villalobos, *Reporte de año: Dinámica poblacional y movimientos del oso negro en el norte de Coahuila, México* (Mexico City: Secretaría de Medio Ambiente y Recursos Naturales, 2005).

21. Leopold, *Wildlife of México.*

22. LeCount and Yarchin, *Black Bear Habitat Use.*

23. Onorato and Hellgren, "Black bear at the border."

24. Onorato and Hellgren, "Black bear at the border"; C. Varas, C. López-Gonzáles, J. Ramirez, P. Krausman, and M. Culver, "Population structure of black bears in northern Mexico," in *Borders, Boundaries, and Time Scales: Proceedings of the Sixth Conference on Research and Resource Management in the Southwest Deserts, Extended Abstracts*, 80–82 (Tucson: Colorado Plateau Research Station, U.S. Geological Survey, 2007). See also Onorato et al., "Phylogeographic patterns within a metapopulation of black bears."

25. Onorato and Hellgren, "Black bear at the border."

26. K. A. Logan and L. L. Sweanor, *Desert Puma: Evolutionary Ecology and Conservation of an Enduring Carnivore* (Washington, D.C.: Island Press, 2001).

27. Schmidly, *Natural History of Texas.*

28. Logan and Sweanor, *Desert Puma.*

29. Schmidly, *Natural History of Texas.*

30. M. E. Pittman, B. P. McKinney, and G. Guzman, *Ecology of the Mountain Lion on Big Bend Ranch State Park in Trans-Pecos Texas* (Austin: Texas Parks and Wildlife Department Press, 2000).

31. K. Hansen, *Cougar, the American Lion* (Flagstaff, Ariz.: Northland, 1992).

32. Beier, "Determining minimum habitat areas and corridors for cougars"; Logan and Sweanor, *Desert Puma.*

33. Logan and Sweanor, *Desert Puma.*

34. Walker et al., "Microsatellite variation in two populations of mountain lions."

35. Roy McBride, independent biologist in Alpine, Texas, personal communication, November 2006.

36. B. H. McRae, P. Beier, L. E. DeWald, L. Y. Huynh, and P. Keim, "Habitat barriers limit gene flow and illuminate historical events in a wide-ranging carnivore, the American puma," *Molecular Ecology* 14 (2005): 1965–77.

37. M. E. Roelke, J. S. Martenson, and S. J. O'Brien, "The consequences of demographic reduction in the endangered Florida panther," *Animal Conservation* 9 (1993): 115–22.

38. D. Brown and C. López-Gonzáles, *Borderland Jaguars* (Salt Lake City: University of Utah Press, 2001), 137–41.

39. Ibid.

40. C. A. López-González and D. E. Brown, "Distribución y estado de conservación del jaguar en el noroeste de México," in *El jaguar en el nuevo milenio*, compiled by R. A. Medellín, C. Chetkiewicz, A. Rabinowitz, K. H. Redford, J. G. Robinson, E. Sanderson, and A. Taber, 379–92 (Mexico City: Fondo de Cultura Económica, Universidad Nacional Autónoma de México, Wildlife Conservation Society, 2001).

41. B. Miller, R. Reading, J. Srittholt, C. Carroll, R. Noss, M. Soule, O. Sanchez, J. Terborgh, D. Brightsmith, T. Cheeseman, and D. Foreman, "Using focal species in the design of nature reserve networks," *Wild Earth* 8 (1998–99): 81–92.

42. M. E. Tewes and D. D. Everett, "Status and distribution of the endangered ocelot and jaguarundi in Texas," in *Cats of the World: Biology, Conservation, and Management*, edited by S. D. Miller and D. D. Everett, 147–58 (Washington, D.C.: National Wildlife Federation, 1986).

43. C. A. López-Gonzáles, D. E. Brown, and J. P. Gallo-Reynoso, "The ocelot *Leopardus pardalis* in north-western Mexico: Ecology, distribution, and conservation status," *Oryx* 37 (2003): 358–64.

44. López-González and Brown, "Distribución y estado de conservación del jaguar"; S. Avila, "Cuatro Gatos Project: Wildlife research and conservation in northwestern Mexico and implications of the border fence," *Wild Field Monitor* 2 (2009):15.

45. K. Hansen, *Bobcat, Master of Survival* (Oxford, U.K.: Oxford University Press, 2007).

46. Ibid.

47. Tony Dee, School of Natural Resources and Environment, University of Arizona, personal communication, July 2007.

48. C. C. Chester, *Conservation across Borders: Biodiversity in an Interdependent World* (Washington, D.C.: Island Press, 2006).

49. M. E. Sunquist and F. Sunquist, "Changing landscapes: Consequences for carnivores," in *Carnivore Conservation*, edited by J. L. Gittleman, S. M. Funk, D. Macdonald, and R. K. Wayne, 399–418 (Cambridge, U.K.: Cambridge University Press, 2001).

1

en Skies over a Closing Border

U.S.–Mexico Efforts to Protect Migratory Birds

Charles C. Chester and Emily D. McGovern

In a Nutshell ──────────────────────────────

- Conservation of migratory birds presents unique challenges due to the number of habitats they utilize, requiring collaborative approaches at local, national, and international levels.

- Protection of migratory bird habitat is critical for all stages of migration: summer breeding ranges, wintering grounds, and stopover sites.

- Four initiatives are representative of conservation efforts spanning the U.S.–Mexico border: the 1936 U.S.–Mexico Convention, the North American Waterfowl Management Plan, Partners in Flight, and the North American Bird Conservation Initiative.

- Although more than twenty bilateral U.S.–Mexico conservation efforts have been aimed at protecting migratory birds, these efforts need greater integration, financial support, and coordination.

Introduction

Since the first half of the twentieth century, conservationists have been working across the U.S.–Mexico border to protect migratory birds. This chapter evaluates these efforts in relation to three areas of inquiry: (1) What is the overall status of bird Species of Conservation Concern that migrate across the U.S.–Mexico border? (2) What has been done on a bilateral basis to protect these migratory birds? (3) How can these transborder efforts become more effective? Although drawing from the literature on conservation status, this chapter focuses more closely on the second issue of bilateral activities. We summarize our findings from research on twenty-one distinct cooperative initiatives on migratory birds that have involved governments or civil society actors, or both, from Mexico and the United States, focusing on four detailed examples.

Although it has proven difficult to assess the effectiveness of these conservation initiatives, we find that the success of future efforts depends mostly on the extent of protection afforded to wintering, breeding, and stopover habitats for migratory birds. Less obvious is the need for further integration and coordination between the myriad international conservation initiatives for migratory birds; there is room for further coordination, but the diversity of initiatives accurately mirrors the complex conservation needs of the vast variety of migratory birds.

"What has been done" to protect migratory birds across the U.S.–Mexico border consists of a potentially confusing array of more than twenty individual (sometimes intersecting) initiatives (see table 7.1). Even knowledgeable conservationists find this array of initiatives labyrinthine and confusing—not to mention potentially redundant and wasteful. Sorting through this expansive web of conservation initiatives constitutes an important step toward more effective conservation of migratory birds. The ultimate purpose of this review is accordingly to help conservationists understand where and how they can better concentrate their efforts (the third area of inquiry).

Foundations of Transborder Conservation

One might reasonably wonder *why* so many transborder initiatives on migratory birds have arisen between the United States and Mexico. In part, they have been created in a piecemeal fashion according to various taxonomic (and pseudotaxonomic) groupings of bird species. They also result in part from the fact that *migratory bird* is a generic descriptor that belies the biological complexity of bird movements across and between continents. Most relevant here is the fact that many migratory birds are not simply flying to and from Mexico and the United States, but to and from other countries in the Americas, including Canada (although it is nonetheless estimated that more than half of North America's migratory birds spend six to eight months per year in Mexico).[1] Consequently, only a few of the initiatives in table 7.1 include solely the United States and Mexico as participants. A more probable cause of all these initiatives is tied to the large number of avian species crossing the border and the resultant complexity in determining their current conservation status. This task is hardly straightforward; it rests on a set of underlying questions that includes: To what degree does the available scientific evidence indicate that migratory

Table 7.1. Selected U.S.–Mexico conservation initiatives for migratory birds.

Year	Cooperative Initiatives	Type
1936	U.S.–Mexico Convention for the Protection of Migratory Birds and Game Mammals	U.S.–Mexico initiatives and treaties
1941	Convention on Nature Protection and Wild Life Preservation in the Western Hemisphere	Western Hemisphere initiatives
1951	North American Flyway Councils (First Formal Proposal)	North American initiatives
1971	Convention on Wetlands of International Importance Especially as Waterfowl Habitat	Multilateral initiatives and treaties
1975	U.S.–Mexico Joint Committee for the Conservation of Wild Flora and Fauna	U.S.–Mexico initiatives and treaties
1985	Important Bird Areas	Multilateral initiatives and treaties
1985	Western Hemisphere Shorebird Reserve Network	Western Hemisphere initiatives
1988	U.S./Mexico/Canada Tripartite Agreement on the Conservation of Wetlands and Their Migratory Birds	North American initiatives
1986	North American Waterfowl Management Plan (NAWMP)	North American initiatives
1989	Partners in Flight	North American initiatives
1994	North American Agreement on Environmental Cooperation and the Commission for Environmental Cooperation	North American initiatives
1994	U.S.–Mexico Wildlife Without Borders Program	U.S.–Mexico initiatives and treaties
1996	Canada/Mexico/U.S. Trilateral Committee for Wildlife and Ecosystem Conservation and Management	North American initiatives
1998	Waterbird Conservation for the Americas	Western Hemisphere initiatives
1999	North American Bird Conservation Initiative	North American initiatives
1999	Non–NAWMP Joint Ventures and the Sonoran Joint Venture	U.S.–Mexico initiatives and treaties
2000	Neotropical Migratory Bird Conservation Act	U.S. federal initiatives
Circa 2000	Park Flight Migratory Bird Program	U.S. federal initiatives
Circa 2003	Western Hemisphere Migratory Species Initiative	Western Hemisphere initiatives
2005	Wings across the Americas	U.S. federal initiatives
2005	Declaration of Intent for the Conservation of North American Birds and Their Habitat	North American initiatives

birds are in decline? Which particular migratory species are in decline? What are the causes of their decline, and which causes are the most significant? How can their decline be most effectively averted? Where should conservation efforts focus geographically?

The North American conservation community has witnessed decades of scientific inquiry and scholarly publications on these questions. These endeavors have revealed much about migratory birds' needs and inspired myriad conservation activities—including the transborder projects described in this chapter. Nonetheless, the status and trends of North America's migratory birds constitute a prominent and long-standing debate within the already contentious battleground of the conservation sciences. The debate lends itself neither to generalization nor to summarization and typically orbits around the extraordinarily difficult empirical challenge of discerning population trends across many different species. Yet with all these caveats, it seems fair to characterize many ornithologists as extremely concerned over the long-term conservation prospects for migratory birds in North America.

Scale and Scope of the Problem

Before turning to examine specific initiatives, we need to clarify and expand our definition of birds that migrate between the United States and Mexico. These birds are known as "Neotropical migratory birds," but not all cross over the U.S.–Mexico land border or the Gulf of Mexico. Numerous species, for instance, make their way to wintering habitat in the Caribbean or South America via Florida or make long traverses over the Atlantic Ocean. But even for those species that do cross the border, the border per se does not constitute a significant obstacle—as it does, for instance, to an earthbound animal such as the Sonoran pronghorn. As discussed more fully later, the problem of transboundary conservation of migratory birds has less to do with the physical border as a biological barrier than with the entire migration range and humans' ability to conceive and coordinate effective conservation responses.

Discussion

Migratory Species' Vulnerabilities

Ornithologists routinely point out that migration constitutes an optimal survival strategy. To appreciate the benefits of migration, one need only juxtapose the fall migration's payoff in tropical calories to the relative dearth of sustenance that nonmigratory birds have evolved to cope with in a temperate, boreal, or arctic winter environment. Furthermore, migratory birds face the same gamut of threats faced by resident species and are no more exposed to them under "normal environmental conditions that have not been adversely affected by humans."[2] But scientists have pointed out that migratory birds *are* particularly vulnerable to the degree that they depend on a geographical chain of resources, any individual link of which can be threatened by the stressors of habitat loss and fragmentation.

The Need for Transboundary Conservation Initiatives

The conservation problem for migratory birds is thus a complex web of threats throughout breeding ranges, wintering ranges, and stopover sites along their migration routes. Because these threats differ from taxa to taxa and from habitat to habitat, the long-term viability of all species of migratory birds requires conservation efforts at multiple scales. Even as conservationists must think and plan in continental and transcontinental scales, the "real work" of conservation must occur at regional and local scales within the varied habitats where migratory species breed, winter, and stop to rest and refuel. This means that both Mexico and the United States must find ways to protect migratory species *within* their own borders.

The fact remains that by crossing an international border, migratory species do raise particular conservation challenges—most notably the impossibility of protection by a single sovereign authority and the concomitant need for international coordination and cooperation. Even if it is not the border per se that constitutes the conservation problem, conservation *solutions* will require transborder cooperation across the U.S.–Mexico divide. These solutions will entail continued exchange of financial, informational, and logistical support and on-the-ground field experience flowing in both directions.

Although it is not widely recognized, Mexican federal laws for the protection of wildlife were established as early as 1894, and by 1932 Mexico had already taken significant steps to protect migratory birds.[3] Nevertheless, enforcement capacity has been sporadic at best, and in comparative terms the United States simply has always had greater financial resources to support migratory bird conservation. A natural result has been the accumulation of a relatively higher capacity for grappling with the practical challenges of protecting migratory birds in the United States. Most of the international initiatives in table 7.1 focus on enhancing support for conservation *in* Mexico for migratory birds.

U.S.-Mexico Cooperative Efforts

Conservationists have long looked across international borders to protect migratory bird species, and perhaps the earliest intimation of U.S.–Mexico cooperation occurred in 1910, when the chief of the U.S. Biological Service suggested the possibility of cooperative conservation efforts to a Mexican government scientist.[4] Soon thereafter, government officials in the United States considered approaching Mexico to propose a treaty for the protection of certain bird species that migrated between the two countries, but this agreement would not occur until 1936 with the U.S.–Mexico Convention for the Protection of Migratory Birds and Game Mammals.

Since that time, conservationists, scientists, and government officials in the two countries have entered into a sizeable—and often confusing—number of bilateral and multilateral conservation initiatives that either focus on migratory birds or include migratory bird conservation as an integral component of their respective missions. Table 7.1 lists these initiatives chronologically according to date of establishment (with a few exceptions due to inexact origination points) to illustrate how they were often either built on or inspired by previous initiatives. It also categorizes them using five general headings: (1) multilateral initiatives and treaties, (2) Western Hemisphere initiatives, (3) North American initiatives, (4) U.S.–Mexico initiatives and treaties, and (5) U.S. federal initiatives.

In addition to demonstrating that transborder cooperation for biodiversity conservation is nothing new, this summary displays the diversity in institutions and approaches to transborder conservation. Because of the varied scales of the initiatives, it also gives lie to the idea that international cooperation must be bound up in formalistic diplomatic protocols. Out of twenty-one initiatives, we have chosen four to portray at some length later in this chapter: the 1936 U.S.–Mexico Convention, the North American Waterfowl Management Plan, Partners in Flight, and the North American Bird Conservation Initiative (NABCI). Although we have chosen these four largely on the grounds of their precedence, influence, and size, it is significant that they are representative of four very different types of transborder initiatives: (1) a traditional bilateral treaty signed between sovereign states, (2) an agreement rooted in cooperation between regulatory government agencies, (3) an active and extensive network of stakeholder groups, and (4) an overarching initiative by a trilateral organization that was established as something of a side effect to regional economic integration (namely, the North American Free Trade Agreement).

U.S.–Mexico Convention for the Protection of Migratory Birds and Game Mammals

In the 1910s, conservationists within the U.S. government sought to ensure federal control over migratory birds by signing an international treaty for their conservation. Although their first prospective partner was Mexico, that country was in the midst of a revolution (from 1910 to 1920), so entering into a treaty with the fractured nation was quickly ruled out as a viable option.[5] The United States subsequently turned to Canada and its more receptive British government (which controlled Canada's foreign relations) to craft an agreement (the 1916 Migratory Bird Treaty).

It was not until two decades later that the United States and Mexico agreed to the 1936 convention. Calling for the two signatories to avoid species extinction through the "rational utilization of migratory birds for the purpose of sport as well as for food, commerce and industry," the convention contained provisions on hunting laws and closed seasons (for birds as well as nests and eggs) and created refuge zones. The United States and Mexico significantly agreed "not to permit the transportation over the American-Mexican border of migratory birds, dead or alive, their parts or products, without a permit of authorization." The 1936 treaty is widely credited with being an essential stepping stone to later collaborative efforts between the United States and Mexico and is considered foundational to subsequent transborder conservation efforts between the two countries.[6]

North American Waterfowl Management Plan and Joint Ventures

Building on decades of coordinated research and conservation work, wildlife officials in Canada and the United States adopted the North American Waterfowl Management Plan (NAWMP) in 1986.[7] In 1994, Mexico formally joined during the plan's first update. Its overarching goal to restore waterfowl populations to levels of the 1970s,[8] NAWMP has been credited with having helped raise U.S.$4.5 billion for waterfowl and wetland protection, and with having protected 6.4 million hectares (15.7 million acres) of wetland habitat.[9] NAWMP quickly achieved recognition as an effective model for international conservation and collaboration and for conducting landscape-level conservation.[10]

One of NAWMP's principal effects on Mexico has been through the establishment of a grant-making program under supporting legislation, the U.S. North American Wetland Conservation Act (NAWCA) of 1989, which aims to support conservation, restoration, and enhancement projects for wetlands and associated upland habitats in the United States, Canada, and Mexico. In addition, projects in Mexico can involve technical training, environmental education and outreach, organizational infrastructure development, and sustainable-use studies.[11]

NAWMP's "demonstrated success" has been attributed to substantial funding, a science-based approach, and a system of partnerships called "Joint Ventures" (JVs).[12] The hallmark of NAWMP's operations, JVs are voluntary, nonregulatory, public-private partnerships involving conservation nongovernmental organizations, businesses, and governments at the federal, state, provincial, and local levels. Although a few JVs focus on either a particular species (e.g., the American Black Duck, *Anas rubripes*) or groups of waterfowl (Arctic geese and sea ducks), most have been "habitat JVs" focused on developing coordinated, science-based, habitat-management programs that address wetland habitat protection in particular regions within North America.[13]

Notably, none of the habitat JVs under NAWMP covers Mexico. However, the success of NAWMP's JV model has led to the development of seven additional JVs outside of NAWMP's purview, two of which focus on the U.S.–Mexico transborder region. Despite its name, the Sonoran Joint Venture (SJV) includes the Mojave Desert, western portions of the Chihuahuan Desert, and a number of other ecosystems.[14] Initiated in 1999, the SJV includes the participation of more than twenty-five agencies and organizations, including the U.S. Fish and Wildlife Service (FWS), the Arizona Game and Fish Department, The Nature Conservancy, Pronatura Noroeste, and the Sonoran state wildlife agency. Under the guidance and direction of a management board, two committees (the Technical Committee and the Communications, Education, and Outreach Committee) conduct the bulk of the SJV's conservation work. The SJV was the first venture to take an "all-bird" approach, as promulgated by the U.S. FWS and promoted under the NABCI (which is discussed later), by expanding its purview far beyond waterfowl.[15] The SJV has been credited with a broad array of on-the-ground conservation collaborations, most of which would not have been able to garner support outside of the JV framework.[16] The second U.S.–Mexico JV is the Rio Grande Joint Venture, which is still in its formative stages.[17]

Partners in Flight

In 1989, the same year that NAWCA was enacted, the Manomet Bird Obser-
vatory (now the Manomet Center for Conservation Sciences) convened a
group of researchers who were concerned about declining populations of
Neotropical migratory birds. That symposium attracted the attention of the
National Fish and Wildlife Foundation, which was interested in replicat-
ing the success of NAWMP for land birds.[18] The foundation sponsored a
second meeting in 1990, out of which came the Neotropical Migratory Bird
Conservation Program, now called Partners in Flight (PIF).[19]

Although PIF originated with a focus on Neotropical migrants, coverage has spread to include a large group of more than four hundred land
bird species from fifty-eight taxonomic families.[20] As the U.S. National
Coordinator for PIF has characterized the program, "Partners in Flight
has expanded from a focus on long-distance migrants that inhabit eastern
deciduous forests to essentially all birds in all habitats."[21] Although PIF's
primary goal is to "keep common birds common," it also aims to protect
"species at risk."[22]

PIF has been described as having a "unique organizational hierarchy—
that is, a relative lack thereof,"[23] and, indeed, the activities that go on under
PIF's aegis are legion. Although NAWMP has served as an institutional role
model, PIF has been a largely U.S.–driven initiative, the "partners" consisting mostly of federal, state, and local government agencies, philanthropic
foundations, professional organizations, conservation groups, industry, the
academic community, and private individuals.[24] However, the International
Working Group under PIF has focused on increasing networking opportunities for the international community, most visibly with the publication of *La
Tangara*, an electronic newsletter aimed at PIF partners in Latin America.
Mexican individuals are involved in various PIF committees and constitute
two of the twelve authors of the 2004 *Partners in Flight North American
Landbird Conservation Plan*, a "continental synthesis of priorities and
objectives that will guide landbird conservation actions at national and
international scales."[25] Although the scope of this major report was limited
to the 448 native land birds that regularly breed in the United States and
Canada, the PIF Science Committee is working with a number of Mexican
partners to expand coverage to another 445 breeding species in Mexico in
an extension to the plan tentatively due out in early 2010 under the title
"Saving Our Shared Birds."[26]

The North American Bird Conservation Initiative

The NABCI was established in 1999 under the aegis of the Commission for Environmental Cooperation (CEC). The CEC is the coordinating body of the North American Agreement on Environmental Cooperation, which came into force in 1994 along with the North American Free Trade Agreement. Up until 2008, NABCI was one of a number of broad initiatives under the CEC's program Conservation of Biodiversity.

NABCI covers all native North American birds—some 1,100 species—be they migratory or not, with the broad mission of ensuring "that populations and habitats of North America's birds are protected, restored and enhanced through coordinated efforts at international, national, regional and local levels guided by sound science and effective management." NABCI has been conceived of as something of a network of networks, a forum linking the many international initiatives listed in table 7.1 with myriad national-level conservation initiatives. Yet it also has an institutional structure that functions through four principal steering committees: the Tri-National Steering Committee plus one committee for each country.[27]

In practice, NABCI focuses on fostering coordination and cooperation across borders and on increasing "the effectiveness of existing and new initiatives" through existing regional partnerships, most importantly the JV programs.[28] Perhaps NABCI's most tangible accomplishment to date has been mapping the continent into Bird Conservation Regions, of which sixty-seven were identified in 2000 (many of the initiatives listed in table 7.1 have revised their approaches to reflect this regional framework).[29] In Mexico, NABCI activities have revolved around the development of the Important Bird Area Program, which is considered the "primary implementation" approach for NABCI in Mexico and is institutionally housed in the country's National Commission for the Knowledge and Use of Biodiversity.[30]

In terms of fostering international collaboration, NABCI was instrumental in the creation of the 2005 Declaration of Intent for the Conservation of North American Birds and Their Habitat, a high-level initiative signed by the interior ministries of Canada, Mexico, and the United States. The declaration calls for the conservation of "native North American birds throughout their ranges and habitats."[31] More generally, NABCI points to the declaration as "a means of increasing the profile and recognition" of the many different initiatives under NABCI's aegis—namely, many of the initiatives listed in table 7.1.[32]

Conclusions

The numerous migratory bird initiatives listed in table 7.1 collectively raise two principal questions. First, to what degree have these bilateral and multilateral initiatives succeeded in protecting North American migratory birds? Second, what needs to happen under (and between) these initiatives in order to maximize migratory bird conservation in the future?

With regard to the first question, there is unfortunately no straightforward accounting of the status of migratory birds. Most likely, the beneficial effects of the twenty-one initiatives individually range from "hardly any" to "considerable but could have been greater." It is safe to aver that these initiatives cumulatively have had a greater effect than had conservationists never bothered to work beyond their national borders. Yet it is difficult to assess the initiatives' effectiveness. Biological systems are extraordinarily complex, and the resources available to monitor the vast number of species and their myriad interactions are inadequate. In regard to the second question, asking what remains to be accomplished, there are two responses— one a stern commandment, the other an indeterminate precaution. The commandment is obvious, but requires endless reiteration: *protect all types of migratory bird habitat.* From the 1936 treaty to the 2005 declaration, a bedrock challenge has been to ensure that migratory birds have places they can breed over summer, survive over winter, and both rest and eat during migration. These challenges do not minimize age-old challenges such as overexploitation or more recently recognized threats such as climate change. Yet meeting any of these challenges will be for naught if we have not protected habitat. The good news is that the conservation community fully recognizes this basic fact and is assiduously attempting to do something in response; the bad news is that this community not only faces an uphill climb in finding the dollars to protect habitat, but also continually has to educate both the public and decision makers about the critical need for habitat protection.

The second response concerns the need for further integration and coordination among the diversity of human institutions working on North American migratory bird conservation. As a major 2007 publication in birding circles noted, although integration is beginning, "it will need to increase to an unprecedented level in the next decades if we are to tackle the challenges before us in the 21st century."[33] Yet caution must be taken here, for although integration seems an obvious solution to the problem of complexity, integration will no doubt raise new challenges. Consider that *migratory bird species* is a catchall phrase, one that glosses over the tremendous diversity both among individual species and among different groups of birds. Because such diversity demands different conservation responses, a simplistic approach to administrative consolidation or centralization of these disparate initiatives could ignore these biological realities. As one biologist described the current status of North American bird conservation to the authors, "While you can always integrate a little more or a little better, we are past the point of diminishing returns."

The success or failure of the broad community of North American bird conservationists will ultimately depend on whether it can engage multiple audiences—ranging from the general public to decision makers—in delivering coherent messages regarding *why* bird conservation in North America matters; *what* North American birds need to survive; *what* resources conservationists will need to get the job done; *how* the conservation community's various partner initiatives are working to protect migratory birds; and *how* a diversity of human institutions can effectively implement on-the-ground conservation.

Overall, conservationists need to remind themselves constantly that even where decision makers do understand the need for conservation, competition for attention in the policy arena is intense. It is a challenge worth reiterating: bird conservationists must always keep in mind that the problem of "migratory bird species" constitutes but a subcategory of biodiversity conservation, which in turn is only one aspect of a broad swath of environmental issues and amounts to only one particular dimension of the complex U.S.–Mexican relationship. If bird conservationists fail to take sociopolitical realities into account in promoting migratory bird conservation, they are unlikely to garner lasting conservation for migratory birds.

Acknowledgments

We benefited from informal conversations with numerous attendees of the 2006 North American Ornithological Conference in Veracruz, Mexico, the 2007 annual conference of the Wildlife Society in Tucson, Arizona, and the 2007 PIF "Tundra to Tropics" Conference in McAllen, Texas. We thank in particular Terrell Rich (PIF's national coordinator at the U.S. FWS) for his extensive comments on a draft of this document, although all errors and matters of judgment remain ours.

Notes

1. M. Arizmendi, L. M. Valdelamar, and H. Berlanga, "Priority setting for bird conservation in Mexico: The role of the Important Bird Areas Program," in *Bird Conservation Implementation and Integration in the Americas: Proceedings of the Third International Partners in Flight Conference, 2002 March 20–24; Asilomar, California,* edited by C. J. Ralph and T. D. Rich, 1256–62, General Technical Report PSW-GTR-191 (Albany, Calif.: Pacific Southwest Research Station, U.S. Forest Service, 2005).

2. P. Berthold, *Bird Migration: A General Survey,* 2d ed. (New York: Oxford University Press, 2001), 180.

3. L. Simonian, *Defending the Land of the Jaguar: A History of Conservation in Mexico,* 1st ed. (Austin: University of Texas Press, 1995), 65, 101–2.

4. K. Dorsey, *The Dawn of Conservation Diplomacy: U.S.–Canadian Wildlife Protection Treaties in the Progressive Era* (Seattle: University of Washington Press, 1998), 196.

5. Ibid., chap. 6 n. 6.

6. U.S. Fish and Wildlife Service (FWS) and General Directorate for Conservation and Ecological Use of Natural Resources of Mexico (DGCEUNR), *Sixty Years of Cooperation Between the United States and Mexico in Biodiversity Conservation (1936–1996)* (Prado Norte, Mexico: U.S. FWS and DGCEUNR, n.d.). The full text of the convention can be found at http://www.fws.gov/le/pdffiles/mexico_Mig_Bird_Treaty.pdf; the quotes come from Article 13.

7. J. D. Nichols, F. A. Johnson, and B. K. Williams, "Managing North American waterfowl in the face of uncertainty," *Annual Review of Ecology and Systematics* 26 (1995): 177–99.

8. E. G. Bolen, "Waterfowl management: Yesterday and tomorrow," *Journal of Wildlife Management* 64 (2) (2000): 329.

9. J. V. Wells, *Birder's Conservation Handbook: 100 North American Birds at Risk* (Princeton, N.J.: Princeton University Press, 2007), 34.

10. J. Wilson, "The Commission for Environmental Cooperation and North American migratory bird conservation: The potential of the NAAEC [North American Agreement on Environmental Cooperation] citizen submission procedure," *Journal of International Wildlife Law and Policy* 6(3) (2003): 205–31; R. Boardman, "Multi-level environmental governance in North America: Migratory birds and biodiversity," in *Bilateral Ecopolitics: Continuity and Change in Canadian-American Environmental Relations,* edited by P. G. Le Prestre and P. J. Stoett, 179–96 (Burlington, Vt.: Ashgate, 2006).

11. PG7 Consultores, S.C., and Faunam, A.C., *Final Report Programmatic Evaluation of the NAWCA Program in México: 1991–2001* (Mexico City: n.p., June 2003), available at http://www.fws.gov/birdhabitat/Grants/NAWCA/files/programmaticevaluationsMX.pdf.

12. J. W. Fitzpatrick, "The AOU and bird conservation: Recommitment to the revolution," *The Auk* 119(4) (2002): 908.

13. U.S. Environmental Protection Agency (EPA), *Bird Conservation Initiatives* (Washington, D.C.: U.S. EPA, March 8, 2006), available at http://www.epa.gov/owow/birds/bird.html, accessed December 6, 2006.

14. Sonoran Joint Venture (SJV), *Sonoran Joint Venture Region* (Tucson, Ariz.: SJV, June 8, 2005), available at http://www.sonoranjv.org/about_us/sjvmap.html, accessed December 15, 2006.

15. North American Waterfowl Management Plan (NAWMP), *2004 Strategic Guidance: Strengthening the Biological Foundation* (Arlington, Va.: NAWMP, U.S. Fish and Wildlife Service, 2004), 19, available at http://www.fws.gov/birdhabitat/NAWMP/files/NAWMP2004.pdf, accessed September 24, 2009.

16. J. P. Cohn, "Joint ventures: A different approach to conservation," *BioScience* 55(10) (2005): 824–27.

17. Texas Parks and Wildlife Department (TPWD), *Rio Grande Joint Venture: An Invitation to Action*, PWD BK W7000-1177 (6/06) (Austin: TPWD, 2006), available at http://www.tpwd.state.tx.us/publications/pwdpubs/media/pwd_bk_w7000_1177.pdf, accessed September 24, 2009.

18. E. Santana C., "A context for bird conservation in México: Challenges and opportunities," in Ralph and Rich, eds., *Bird Conservation Implementation and Integration in the Americas*, 15–25.

19. J. Faaborg, "Partners in Flight North American Landbird Conservation Plan," *The Auk* 122(1) (2005): 373–75.

20. T. D. Rich, C. J. Beardmore, H. Berlanga, P. J. Blancher, M. S. W. Bradstreet, G. S. Butcher, D. W. Demarest, E. H. Dunn, W. C. Hunter, E. E. Iñigo-Elias, J. A. Kennedy, A. M. Martell, A. O. Panjabi, D. N. Pashley, K. V. Rosenberg, C. M. Rustay, J. S. Wendt, and T. C. Will, *Partners in Flight North American Landbird Conservation Plan* (Ithaca, N.Y.: Cornell Lab of Ornithology, March 2005), available at http://www.partnersinflight.org/cont%5Fplan, accessed September 24, 2009.

21. T. D. Rich, "Partners in Flight: Working for bird conservation implementation and integration in the Western Hemisphere," in Ralph and Rich, eds., *Bird Conservation Implementation and Integration in the Americas*, 5.

22. Partners in Flight–U.S. (PIF–U.S.), *What Is Partners in Flight?* (N.p.: PIF–U.S., November 28, 2006), available at http://www.partnersinflight.org/description.cfm, accessed July 7, 2006.

23. C. J. Ralph and T. D. Rich, "The state of the art and the state of science: Partners in Flight in the 21st century," in Ralph and Rich, eds., *Bird Conservation Implementation and Integration in the Americas*, 1.

24. PIF–U.S., *What Is Partners in Flight?*

25. Rich et al., *Partners in Flight North American Landbird Conservation Plan*.

26. *Partners in Flight Newsletter* (July 2009), available at http://www.partnersinflight.org/pubs/pifnews.cfm, accessed September 24, 2009.

27. North American Bird Conservation Initiative (NABCI), *About NABCI* (n.d.), available at http://www.nabci.net/international/english/about.htm, accessed November 30, 2006.

28. Ibid.

29. U.S. NABCI Committee, *Bird Conservation Region Descriptions: A Supplement to the North American Bird Conservation Initiative Bird Conservation Regions Map* (Arlington, Va.: U.S. NABCI Committee, September 2000), available at http://www.nabci-us.org/aboutnabci/bcrdescrip.pdf, accessed September 24, 2009.

30. Commission for Environmental Cooperation (CEC) Council, *Council Resolution 99-03: North American Bird Conservation Initiative*, C/99-00/RES/03/Rev.8 (Montreal: CEC Council, June 28, 1999), available at http://www.cec.org/files/PDF/ABOUTUS/Council_resolution99-03e_EN.pdf, accessed September 25, 2009..

31. Department of the Environment of Canada (DEC), Secretariat of the Environment and Natural Resources of the United Mexican States (SEMARNAT), and U.S. Department of the Interior (DOI), *Declaration of Intent for the Conservation of North American Birds and Their Habitat Between the Department of the Environment of Canada, the Department of the Interior of the United States of America, and the Secretariat of the Environment and Natural Resources of the United Mexican States* (Ottawa, Mexico City, and Washington, D.C.: DEC, SEMARNAT, U.S. DOI, 2005), available at http://www.nabci-us.org/aboutnabci/NABCIFINALDOI.pdf, accessed September 25, 2009.

32. NABCI, "Questions and answers: North American Bird Conservation Initiative (NABCI) Declaration of Intent for the Conservation of North American Birds and Their Habitat," May 25, 2005, available at http://www.nabci-us.org/aboutnabci/Q&A-NABCI-DOI.pdf, accessed September 24, 2009.

33. Wells, *Birder's Conservation Handbook*, 36.

Chapter 8

Mexican Wolf Recovery

Insights from Transboundary Stakeholders

José F. Bernal Stoopen, Jane M. Packard, and Richard Reading

In a Nutshell

- To prevent extinction of endangered species in transboundary ecosystems, stakeholders in the United States and Mexico must coordinate their work.

- For many endangered species of the border region, issues such as insufficient suitable habitat and low genetic diversity can best be solved by cross-border collaboration among stakeholders.

- Regional recovery efforts should be coordinated by binational working groups for each transboundary ecosystem that provides habitat for several species of concern.

- The authors studied participants from both countries in recovery efforts for wolves and other species to document their collaboration priorities and practices.

- Priorities for achieving effective collaboration are: sufficient fund appropriations, multilevel government agency coordination, equitable U.S.–Mexico participation, personnel continuity, cultural exchange, and the establishment of binational working groups.

Introduction

More than a dozen species are at risk of decline or extinction in the ecosystems that cross the U.S.–Mexico border. Recovery programs aimed at reversing extinction have engaged diverse stakeholders in both countries. Transboundary conservation must address the problem of coordinating relatively flexible local responses within the context of relatively inflexible regional, national, and international politics. In working across the international border to protect species at risk, decision makers need information about the actions most likely to benefit stakeholders, which include governmental agencies, landowners, captive-breeding centers, scientists, and nongovernmental organizations (NGOs) engaged in species recovery efforts.[1]

This chapter presents the results of a needs assessment of conservation practitioners working on several species at risk in transboundary ecosystems (listed in table 8.1), including an in-depth case study of recovery efforts for the Mexican wolf *(Canis lupus baileyi)*.[2] First, we provide background information on the Mexican wolf. Next, we briefly describe our study methods and summarize the results. Then we examine the most pressing stakeholder concerns about the conservation of transboundary species and identify priorities for improving cross-border cooperation. We conclude with recommendations to establish binational working groups to coordinate endangered species recovery within each transboundary ecosystem.

Background

The Mexican wolf is a charismatic and wide-ranging subspecies of the gray wolf (fig. 8.1). Its former range crossed the international border in the arid high-elevation mountains and desert ecosystems. Over the past century, however, agricultural expansion policies have led to local extermination of predators such as wolves due to conflicts with humans and livestock. Between 1910 and 1925, U.S. federal trappers reported killing more than nine hundred wolves in Arizona and New Mexico.[3] All but eliminated by 1970,[4] the last wolves trapped in Mexico between 1977 and 1980 were the founders of a captive-breeding population. Mexican wolves have become a symbol for stakeholders who believe that international collaboration can overcome the mismatch between the broad transboundary ranges of more than a dozen endangered species and limited jurisdictions of state and federal government agencies.[5]

Approaches to Recovery of the Mexican Wolf

Because wolves are considered a keystone species, a "systems approach" connecting a broad network of sites is more likely to be successful than a "Band-Aid approach" to conservation of individuals in small, disconnected protected areas.[6] No single protected area in the border region today is sufficiently large to support a viable breeding population of wolves.

A widely accepted metric for successful recovery of Mexican wolves is three connected core areas with at least one hundred wolves each. To find enough prey, the typical wolf family group (two parents, one subadult, and two pups) is likely to range over 400–520 square kilometers (km²) (248–323

Table 8.1. Distribution of selected species at risk in transboundary ecosystems, coded by level of biological concern: (a) species present; (b) species absent, but habitat present; and (c) species extirpated, and former habitat degraded.

Species at risk	Mexican border states						U.S. border states			
	BCN	SON	CHIH	COAH	NL	TAMPS	CA	AZ	NM	TX
—Both Eastern and Western Ecosystems—										
Mexican wolf (*Canis lupus baileyi*)		b	c	c				a	a	c
Jaguar (*Panthera onca*)		a	a		a	a	c	c	c	
Northern Aplomado Falcon (*Falco femoralis septentrionalis*)		a				a		a	a	a
—Primarily Eastern Ecosystems—										
Jaguarundi (*Herpailurus yaguarondi*)						a				a
Kemp's ridley sea turtle (*Lepidochelys kempii*)						a				a
Maroon-fronted Parrot (*Rhynchopsitta terrisi*)						a				a
Mexican long-nosed bat (*Leptonycteris nivalis*)				a	a	a		a	a	a
Mexican prairie dog (*Cynomis mexicanus*)				a	a					
Ocelot (*Leopardus pardalis*)						a				a
—Primarily Western Ecosystems—										
Black-footed ferret (*Mustela nigripes*)		a						a		
Black-tailed prairie dog (*Cynomys ludovicianus*)		a						a	a	a
Imperial Woodpecker (*Campephilus imperialis*)		c	c					c	c	
Mexican grizzly bear (*Ursus arctos nelsoni*)		c	c					c	c	
Mexican Spotted Owl (*Strix occidentalis lucida*)		a	a					a	a	
Pronghorn antelope (*Antilocapra americana peninsula; A. a. sonoriensis; A. a. mexicana*)	a	a	a					a	a	a
Thick-billed Parrot (*Rhynchopsitta pachyrhyncha*)		a	a					a	a	
Vaquita porpoise (*Phocoena sinus*)	a	a					a	a		

Figure 8.1. Mexican wolf in the Chapultepec Zoo in Mexico City, awaiting reintroduction into the wild. Photograph by Antonio Pastrana.

square miles [mi^2]) per year. In a study in the Blue Range Wolf Recovery Area (17,775 km^2 [11,045 mi^2]), consisting of national forest lands in Arizona and New Mexico, dispersing Mexican wolves traveled 70 to 90 km (43–56 mi) before settling into a new area. In a study farther north, typical dispersal distances for gray wolves were 750–1,500 km (466–932 mi).[7]

In 2006, the known population of Mexican wolves consisted of fifty to sixty individuals in nine packs within the reintroduced Blue Range population and about three hundred individuals in forty-eight captive facilities in Mexico and the United States.[8] Although some field biologists believe Mexican wolves may be hidden in remote mountains in Mexico, there is not sufficient evidence to suggest a viable wild breeding population in Mexico.[9] Plans for additional reintroduction sites in Mexico and the United States are under discussion, although opposition by the livestock industry fuels controversy and delays the decision-making process.[10]

U.S. and Mexican Roles in Recovery

The Mexican wolf was officially listed as endangered in the United States in 1976 and in Mexico in 1994. To start the captive-breeding program under jurisdiction of the U.S. Fish and Wildlife Service, Roy McBride captured five wolves in Mexico.[11] Compared to Mexico, the U.S. legal mandate and public pressure to conserve wolves are strong, mobilizing more resources for wolf recovery. However, the majority of potential habitat is in Mexico. In each country, federal dollars were expended on both the extinction (via wolf eradication and agricultural expansion policies) and the recovery of Mexican wolf populations. These seemingly contradictory policies are still debated among agencies at both state and federal levels.

Small, captive populations with high levels of inbreeding may suffer from low reproductive rates, abnormal sperm, high juvenile mortality, and genetic disease.[12] A U.S. federal decision in 1987 that the captive-breeding population of Mexican wolves was doomed to genetic failure was reversed in 1990 by a lawsuit that triggered a series of new actions by the U.S. Mexican Wolf Recovery Team.[13] Prior to recent genetic studies, only the wild-caught wolves were recognized as a "pure" bloodline. They were identified as the "McBride" lineage.[14] Descended from only three individuals, the McBride lineage is small in body size, and litter size has been smaller than expected. Anomalies in both traits raise concern about the genetic health of small captive populations, although it is difficult to distinguish the genetic influence from environmental influences.

Thanks to modern technology and binational collaboration, genetic studies provided enough new scientific evidence to integrate two other bloodlines into the recovery program.[15] Genetic diversity was enhanced by four additional founders from the "San Juan" and "Ghost Ranch" bloodlines, respectively maintained in Mexico and the United States. Although inbreeding was high while the San Juan and Ghost Ranch lineages were isolated, the network of facilities managing the McBride lineage used computer-based management techniques to minimize it. Descendents of these three lineages now interbreed in the reintroduced population in the Blue Range.[16] The subsequent increase in litter size has been dubbed a "genetic rescue."[17]

The Human Dimensions of Mexican Wolf Recovery

When we started our research on stakeholders in 1995, much more was known about the biological than the human dimensions of Mexican wolf recovery.[18] An informal technical advisory group in Mexico facilitated communication within the network of captive-breeding facilities and with the formal U.S. Recovery Team. The captive-breeding effort was managed in Mexico according to the policies established by the Species Survival Plan of the Association of Zoos and Aquariums. Representatives of cooperating

Figure 8.2. Exchange visits between counterparts at Mexican and U.S. captive-breeding facilities are an effective means of building trust, transferring technical expertise, and facilitating cooperation among key actors. Photograph by Jane M. Packard.

institutions met annually to make decisions about which wolves to breed and how to transfer them among institutions (fig. 8.2). For instance, when in the mid-1980s wolves from the McBride lineage filled all available space in U.S. facilities, three breeding pairs were transferred to Mexican facilities. Mexican wolf recovery efforts have been characterized by the strong motivation of a handful of individuals, a dedicated volunteer effort, and private resources donated to augment limited funding from the two federal governments.

Study

Design and Methods

In problems shared by multiple stakeholders, no single agency may have the authority to make decisions that would bind other participating organizations. In this study,[19] we examined the following premise: to the extent that multiple conservation stakeholders are willing to discuss their perspectives on the costs and benefits of species recovery options, collaboration will be more likely. As a result of collaboration, innovative solutions will emerge for mutually beneficial plans of action that integrate local and national interests.

We could find no previously published studies of collaboration on endangered species crossing international borders, so our approach focused on understanding the system through needs assessment and a conceptual model.[20] Our needs assessment included three phases. The first consisted of in-depth semistructured interviews with 44 Mexican wolf stakeholders. In the second, we used a hierarchical thematic analysis[21] of the interviews to identify five major clusters of needs and concerns. Third, we surveyed 250 stakeholders working on multiple endangered species in the border area, including Mexican wolves and others (see table 8.1), asking participants to rank the relative importance of the issues within each cluster. Participants in Mexican wolf recovery were included in the broader survey of other species recovery programs. We integrated results from the in-depth interviews and broader surveys to analyze how social factors influenced stakeholders' priorities about recovery decisions and binational collaboration.[22]

Results

The interviewees from Mexican wolf recovery programs in the United States and Mexico spoke about needs that we categorized within five issue clusters: resources, coordinated binational projects, organizations, culture, and people (interpersonal skills). Interviewees spoke about these issue clusters as interrelated. For example, without a sound base of intercultural understanding, key actors were unlikely to implement binational projects successfully. Likewise, the resources required to support binational collaboration were unlikely to be mobilized without clearly defined roles for organizations on both sides of the border.

The highest priority needs from all clusters were: increased funding, managed with accountability; coordination of federal, state, and local efforts; equitable participation by diverse stakeholders from both nations in project design and implementation; continuity of personnel; and exchange visits to facilitate understanding of diverse cultural perspectives. For 70 percent of the twenty-six need statements, respondents' rankings did not vary significantly by participants' national origin or their inclusion in a particular species recovery program (except as noted in table 8.2). Later in the chapter, we examine each of the issue clusters developed from the wolf program interviews and the multispecies ranking of needs.

Table 8.2. Results of needs assessment: "Within each category, how would you rank the priority of needs over the next three years?"

Issue cluster	Need	Priority Index[a]
Resources	Funding increase	81
	Funding management	71
	Information exchange	68
	New information[b]	66
	Skills training[b]	65
	Technology transfer	54
Coordinated projects	Project design	86
	Project management	73
	Project review	54
	Balance of captive/field effort[c]	49
	National autonomy	41
Organizations	Coordination: federal/state/local	75
	Institutional continuity	72
	Balance of government and nongovernmental organizations	53
	Formal procedures	46
	Decentralization of decision making[b]	45
Culture	Exchange visits	79
	Trust/reciprocity	71
	Bilingual skills	53
	Intercultural skills[e]	49
People (interpersonal skills)	Communication skills	84
	Continuity of participants	73
	Understanding diverse perspectives[c]	73
	Leadership skills	64
	Personal interaction skills[d]	60
	Negotiation skills[e]	52

[a] The Priority Index was based on weighted first through last place rankings within each cluster, standardized within each cluster on a scale of 1 to 100. (See J. F. Bernal Stoopen, "Binational collaboration in endangered species recovery efforts: A case study of the Mexican wolf," Ph.D. diss., Texas A&M University, College Station, 2004.)

[b] Mexican wolf recovery participants ranked item at a significantly lower priority than did participants of other programs (t-test, $P < 0.05$).

[c] Mexican wolf recovery participants ranked item at a significantly higher priority than did participants of other programs (t-test, $P < 0.05$).

[d] Mexican respondents ranked item at a significantly higher priority than did U.S. respondents (t-test, $P < 0.05$).

[e] Mexican respondents ranked item at a significantly lower priority than did U.S. respondents (t-test, $P < 0.05$).

Resources

Respondents from all species programs described funding as the most important element to overcoming barriers to collaboration. In interviews with participants in wolf recovery, a lack of financial resources was repeatedly described as a barrier to basic communication among key actors. Where the *will* has existed to communicate across institutional and national boundaries, unequal distribution of resources has too often blocked the *way* to communicate.

From both sides of the international border, wolf program interviewees perceived a problem with inequitable distribution of economic resources between programs in the two countries in favor of the United States. One interviewee said, "If the level is not balanced . . . this situation will never allow an authentic effort. . . . [T]he day both countries can contribute equally, conservation will be working much better."

Other wolf program interviewees spoke about the practical implications of unmet funding needs: "I think part of the main problem . . . was money and being able to get the individuals funded." In particular, smaller institutions were less likely to have funds for international communication. A participant from one such institution said, "I do not have enough resources to send faxes or to send letters. . . . I have to use a public telephone and stand in line so that I can use it." Interviewees also explained that although smaller institutions were more flexible in decision making and problem solving, their "hands were tied" because they did not possess the resources needed to gain knowledge and implement decisions that might benefit the wider network of recovery participants.

In the quantitative study results, survey respondents from both wolf and nonwolf programs identified "funding increase" as the first-priority, resource-related need (table 8.2). They ranked "technology transfer" as the lowest priority. More than 41 percent of respondents chose "funding management" as the first or second priority. "Skills training" was ranked higher by Mexicans than by Americans. Participants in the Mexican wolf program ranked the need for "new information" significantly lower than did participants in other recovery efforts.

Coordinated Projects

Overall, respondents in both parts of the study indicated a need for clear definition of goals, objectives, tasks, and responsibilities for the participating organizations in both countries. Respondents identified the mobilization of resources and administrative commitment as variables that lead to effective action.

Some wolf recovery program interviewees identified the development of a truly binational plan to coordinate recovery efforts as key to binational collaboration. Interviewees from Mexico tended to use the term *binational program*, whereas those from the United States used the term *recovery plan*. Both terms refer to an official written document that participants would prepare and that the governments of both countries would approve.

Interviewees provided insights on why they perceived a project design that would balance participation from both sides of the border as a high priority for recovery efforts. Some described a shortage of transboundary coordination and reasons for it: "Simply because of the magnitude of the effort that we're trying to accomplish here in the U.S., [we] haven't had very much time to devote to the binational aspect of the effort." Thinking ahead, one interviewee stated, "What it is missing is a project that really identifies what the United States and Mexico need to do, who are the actors in Mexico and who are the actors in the United States, what is their responsibility and what is our responsibility." Additional benefits of a coordinated binational approach to project design and management include: better coordination of people from many disciplines, clearer assignment of responsibilities, the use of documented priorities to guide budget processes within federal institutions, and the ability to leverage decisions within and between agencies.

Some interviewees participating in the Mexican wolf program supported the balancing of input from stakeholders working on the captive breeding and field components (i.e., conservation in the wild). They perceived the U.S. emphasis on ex situ captive breeding as different from the Mexican emphasis on in situ management in the wild. Ironically, U.S. public support for Noah's Ark–style captive breeding has provided leverage for key actors to continue Mexican wolf recovery efforts even when support for expensive field studies has dwindled. Priorities for field studies of wolves in Mexico were swamped by the initiatives spearheaded by private captive-breeding organizations within the structure of the Species Survival Plan.

A trade-off between captive and field efforts was apparently not as salient in recovery efforts for other transboundary endangered species, some of which are easier to maintain in captivity or are less charismatic than wolves. In the needs survey, we investigated whether perceived needs related more to project design or to project management and review (table 8.2). Eighty-three percent of survey respondents ranked "project design" as the first or second priority in this case. The vast majority of respondents

(74 percent) agreed that "project management" was the second or third priority. "National autonomy" was ranked low (fourth or fifth) by 72 percent of respondents, and we found no significant effect of nationality on this or other needs ranked within the project issue cluster.

Organizations

Survey respondents and Mexican wolf program interviewees agreed that better continuity and coordination among local, state, and federal agencies were high-priority needs to facilitate transboundary endangered species recovery (table 8.2). One interviewee illustrated this point, saying, "I dealt with seven or eight different directors in six or seven years. There was no continuity." One interviewee from an NGO proposed that stable nongovernmental institutions, such as zoos, universities, and conservation organizations, might compensate for the lack of continuity within certain government organizations: "I first began stimulating [communication] by inviting individuals from Mexican wolf facilities . . . to come to . . . discuss the issues and to be involved in how we manage the Mexican wolf. . . . Through that process, we developed [and] kept communication going between both countries." In this instance, NGOs facilitated informal communication when a higher-level decision to halt the captive-breeding program of Mexican wolves blocked formal channels. However, participants in other endangered species programs did not echo this perception.

We found that nationality had no significant effect on priority rankings within the organizations cluster (table 8.2). Most survey respondents (60 percent) ranked "coordination of federal/state/local" organizations as their first or second priority. Similarly, 62 percent of respondents ranked "institutional continuity" first or second. "Formal procedures" and a "balance of governmental and nongovernmental organizations" received lower priority rankings. Overall, 61 percent of respondents rated the "decentralization of decision making" as a relatively low priority (fourth or fifth). Participants in the Mexican wolf program ranked "decentralization" significantly lower than did participants in recovery efforts for other species.

Culture

In the context of this study, we use the term *culture* to refer to the tacit understandings that result from shared experiences. Understanding is tacit when members of one group understand each other without explicitly talking about an issue, whereas outsiders will express the need for explanation. When we refer to cultural differences in this study, we mean gaps in communication between captive-breeding practitioners and field researchers as well as between people who reside on different sides of the border.

Interviewees from the border region said that they found it much easier to coordinate efforts at the local level because their families had crossed the border for generations. They believed they understood the culture, language, and methods of reciprocal exchange better than agents of state and federal governments who were from distant locations: "[Collaborating organizations] need people along the border who understand the culture and preferably the language because it's very difficult if you need an interpreter all the time."

Interviewees explained the need for exchange visits to improve trust and reciprocity in a historical context that has at times included paternalistic behavior by the United States toward Mexico. Further, they said, visits must not be superficial exchanges—visiting border towns was not a meaningful intercultural experience. In contrast, visiting workplaces in the heartland of each nation for the purpose of understanding the reasons for differing perspectives, developing personal rapport, and exploring meaningful mutual exchange were believed to be more beneficial. Examples of meaningful exchange included informal gatherings as well as workshops and scientific meetings.

A majority of survey respondents (78 percent) ranked "exchange visits" as the first- or second-highest priority for addressing cultural issues, with no significant difference by nation of origin or recovery program (table 8.2). These results were consistent with the high priority respondents gave in other clusters to developing communication skills, understanding diverse perspectives, coordinating efforts across agency levels, and designing truly binational programs. The second-highest-ranked cultural need was "trust and reciprocity," which 65 percent of respondents agreed was the first or second priority. Respondents (74 percent) also agreed that the need for improving intercultural communication skills should receive the lowest- or second-lowest-priority rank. Overall, U.S. participants tended to rank this need higher than Mexicans.

People (Interpersonal Skills)

Interviewees provided insights into what they meant by the need for personal communication skills as distinct from bilingual skills. They suggested that language barriers might be overcome if both parties were highly committed to recovery efforts. One interviewee from Mexico asserted: "When there is a common interest, even when none of the parties can speak a word of the other language, the objective is the same." Conversely, even good language skills were insufficient if one key actor neglected to make the extra effort to understand diverse perspectives.

Interviewees described the need for personal rapport to facilitate interactions between institutions as a way of moving forward to meet project goals even when mistrust between government agencies inhibited collaboration. In this context, some interviewees explained that just one well-trained cross-cultural person in the international office of one agency was insufficient. The people we interviewed attributed the success of exchange visits in facilitating binational collaboration to the personal rapport and trust that developed as participants at local and regional levels worked together to accomplish tasks, such as exchanging breeding animals in the captive wolf program: "I think . . . there was not a question [of trust on a personal level], but there has always been the question of trust at the level of governments."

"Communication skills" emerged as a top priority within the people cluster (table 8.2), independent of nationality or program. A majority of respondents (58 percent) ranked this need as their first or second priority, and only 5 percent ranked it as their lowest priority (fifth or sixth). Based on our priority index (table 8.2), survey participants equally ranked improving "understanding [of] different perspectives" and promoting "continuity among program participants." Whereas 43 percent ranked continuity as high priority, 25 to 28 percent ranked these needs as lowest priority. Participants in the Mexican wolf program tended to rank "understanding" higher than participants in recovery efforts for other species. Overall, respondents considered "leadership skills" an intermediate priority, with no significant effect of nation or program. Lowest priority was assigned to "personal interaction skills" and "negotiation skills."

Possible Models

Y2Y, an initiative led by NGOs in the Rocky Mountains on the U.S. border with Canada, was established with substantial representation from boards appointed separately in Canada and the United States.[23] Although its huge scale made it vulnerable to critique and misrepresentation by detractors, this initiative may serve as a working model for future initiatives on the U.S.–Mexican border, with some important modifications (such as addressing uncertainties from climate change, illegal narcotics traffic, and water allocation).

A more localized place-based approach, as undertaken by the International Sonoran Desert Alliance, would unlikely meet the habitat needs of wide-ranging species such as Mexican wolves. For these reasons, we emphasize the linking of local and regional initiatives, which would facilitate both flexibility and continuity in transboundary landscapes. José Bernal Stoopen previously recommended an informal approach for the U.S.–Mexican border, matching formal government planning with informal capacity building using independent consultants to maintain the flexibility of small working groups while enhancing information flow among groups.[24]

Incorporating ideas developed by others,[25] we recommend coordinating species-specific, binational working groups as subcommittees under the umbrella of larger, ecosystem working groups that coordinate planning for sets of species found within similar landscapes. Representatives of each country's federal government ideally would meet annually with representatives of different stakeholder groups to hear their perspectives and integrate their needs and recommendations into an adaptive approach to implementing plans. Utilizing the services of experienced consultants, NGOs should play a more active role in acknowledging, understanding, and integrating all stakeholders' perspectives.

In our research, most participants in recovery programs for transboundary species agreed with respect to the highest-priority needs.[26] To meet these needs, we recommend an approach that embeds multiple species within ecosystems to create a binational working group for each of the thirteen transboundary endangered species. A working group for each ecosystem should also address the cross-cutting issues associated with specific endangered species.

Each binational working group should be set up to minimize hierarchical structure, facilitate information exchange, embrace innovation, seek consensus, represent stakeholders from both countries, and encourage better cross-cultural communication.[27] We extend these recommendations to any landscape-level working groups that might form, considering lessons learned in other transboundary initiatives.[28] Binational working groups should not only provide the best biological recommendations for species recovery but also consider pertinent social, organizational, economic, political, and cultural issues.[29] They should ideally remain "task oriented," thereby maximizing trust and reciprocity established via exchange visits, which would in turn minimize the issues of power and authority that can derail decision making.

Conclusions

Through qualitative and quantitative research among stakeholders in species recovery efforts along the international border between Mexico and the United States, we identified priorities for decision makers as they weigh actions that would benefit practitioners working toward biodiversity conservation in the transboundary region. These priorities include increased funding, managed with accountability; continuity and coordination of federal, state, and local efforts; equitable binational participation in project design and implementation; exchange visits to facilitate trust and reciprocity; and improved communication skills and continuity of personnel to foster better understanding of the various stakeholders' diverse perspectives.

We found high levels of agreement on top priorities that were consistent across the international border. These results offer insight that will, we hope, improve conservation success rates in important transboundary ecosystems and across landscapes.

Based on this needs assessment, we recommend establishing binational working groups for each transboundary ecosystem providing habitat for endangered species. Doing so would optimize the use of scarce resources, while facilitating cross-fertilization of ideas and coordination of local, state, and federal efforts.

Overall, this study revealed the value of key actors who are motivated to fill in the gaps between the formal procedures of government agencies. Social dimensions must be considered in conjunction with biological issues in a coordinated, interdisciplinary approach to recovering endangered species in the ecosystems that cross international and state boundaries.

Acknowledgments

We thank all participants in this study. In addition, we acknowledge Wendy Brown, Ernesto Enkerlin, Fred Koontz, Laura López-Hoffman, Emily McGovern, Brian Miller, Carina Miller, David Parsons, Peter Siminski, Edward Spevak, and Priscilla Weeks. We are also grateful for support from a Consejo Nacional de Ciencia y Tecnología (National Science and Technology Council)/Fullbright Fellowship, Mexico's Secretaría de Medio Ambiente y Recursos Naturales (Secretariat of Environment and Natural Resources),

Wildlife Conservation Society, Lincoln Park Zoo Neotropical Fund, Earth Promise Fund of Fossil Rim Wildlife Center, Denver Zoological Foundation, Department of Wildlife and Fisheries Sciences at Texas A&M University, and National Science Foundation Grant no. 0551832. These institutions are not to be held accountable for the content of this chapter.

Notes

1. Although there are many nodes of consensus between different stakeholders, fundamental differences in perspective also exist. For instance, certain key actors have pivotal experience in multiple endangered species programs. Others work within organizations that are more isolated and more narrowly focused on one species.

2. The full study, including methods and interpretation of results, is documented in more detail in J. F. Bernal Stoopen, "Binational collaboration in endangered species recovery efforts: A case study of the Mexican wolf," Ph.D. diss., Texas A&M University, College Station, 2004.

3. D. Brown, *The Wolf in the Southwest: The Making of an Endangered Species* (Tucson: University of Arizona Press, 1983).

4. Ibid.; J. C. Burbank, *Vanishing Lobo: The Mexican Wolf and the Southwest* (Boulder, Colo.: Johnson, 1990); B. Holaday, *Return of the Mexican Gray Wolf: Back to the Blue* (Tucson: University of Arizona Press, 2003).

5. Holaday, *Return of the Mexican Gray Wolf.*

6. G. Ceballos, P. Rodriguez, and R. A. Medellín, "Assessing conservation priorities in megadiverse Mexico: Mammalian diversity, endemicity, and endangerment," *Ecological Applications* 8 (1998): 8–17; C. Carroll, M. K. Phillips, C. A. López-González, and N. H. Schumaker, "Defining recovery goals and strategies for endangered species: The wolf as a case study," *Bioscience* 56 (2006): 25–37.

7. Interagency Field Team (IFT), *Mexican Wolf Blue Range Reintroduction Project 5-Year Review: Technical Component* (Albuquerque: Arizona Game and Fish Department, New Mexico Department of Game and Fish, U.S. Department of Agriculture APHIS and Forest Service, Wildlife Services, U.S. Fish and Wildlife Service, and White Mountain Apache Tribe, 2004).

8. R. J. Fredrickson, P. Siminski, M. Woolf, and P. W. Hedrick, "Genetic rescue and inbreeding depression in Mexican wolves," *Proceedings of the Royal Society B, Biological Sciences* 274 (2007): 2365–71; P. W. Hedrick and R. J. Fredrickson, "Captive breeding and the reintroduction of Mexican and red wolves," *Molecular Ecology* 17 (2008): 344–50.

9. Fredrickson et al., "Genetic rescue and inbreeding depression in Mexican wolves."

10. Agricultural producers raised concerns that livestock operations in Texas, Sonora, California, and Colorado are within potential colonization distance of the Blue Range wolves. See A. Povilitis, D. R. Parsons, M. J. Robinson, and C. D. Becker, "The bureaucratically imperiled Mexican wolf," *Conservation Biology* 20 (2006): 942–45, and E. A. Fitzgerald, "Lobo returns from limbo: New Mexico Cattlegrowers

Ass'n. v.s. U.S.A. Fish and Wildlife Service," *Natural Resources Journal* 46 (2006): 9–64. See also Holaday, *Return of the Mexican Gray Wolf.*

11. Fredrickson et al., "Genetic rescue and inbreeding depression in Mexican wolves"; Hedrick and Fredrickson, "Captive breeding and the reintroduction of Mexican and red wolves."

12. Hedrick and Fredrickson, "Captive breeding and the reintroduction of Mexican and red wolves"; C. Asa, P. Miller, M. Agnew, J. A. R. Rebolledo, S. L. Lindsey, M. Callahan, and K. Bauman, "Relationship of inbreeding with sperm quality and reproductive success in Mexican gray wolves," *Animal Conservation* 10 (2007): 326–31.

13. U.S. Fish and Wildlife Service (FWS) appointed a Mexican Wolf Recovery Team in 1979. See U.S. FWS, *Mexican Wolf Recovery Plan* (Albuquerque: U.S. FWS, 1982). In 1991, U.S. FWS established a new recovery team and hired a full-time recovery coordinator to manage the captive population for reintroduction, to identify potential reintroduction sites in the United States, and to search for evidence of wild wolves in northern Mexico. See D. R. Parsons and J. E. Nicholopolous, "An update of the status of the Mexican wolf recovery program," in *Ecology and Conservation of Wolves in a Changing World,* edited by L. N. Carbyn, S. H. Fritts, and D. R. Seip, 141–46 (Edmonton, Canada: University of Alberta Press, 1995).

14. IFT, *Mexican Wolf Blue Range Reintroduction Project 5-Year Review;* Fredrickson et al., "Genetic rescue and inbreeding depression in Mexican wolves."

15. Hedrick and Fredrickson, "Captive breeding and the reintroduction of Mexican and red wolves."

16. IFT, *Mexican Wolf Blue Range Reintroduction Project 5-Year Review.*

17. Fredrickson et al., "Genetic rescue and inbreeding depression in Mexican wolves."

18. Bernal Stoopen, "Binational collaboration in endangered species recovery efforts."

19. Ibid.

20. Ibid.

21. On this method, see T. R. Peterson, K. Witte, E. Enkerlin-Hoeflich, L. Espericueta, J. T. Lora, N. J. Florey, N. T. Loughran, and R. Stuart, "Using informant directed interviews to discover risk orientation: How formative evaluations based in interpretive analysis can improve persuasive safety campaigns," *Journal of Applied Communication Research* 22 (1994): 199–215.

22. In the results of the survey, potentially confounding effects of demographic variables were documented for national origin, work experience, education, and gender. Three variables were significantly affected by national origin: current position, age, and degree of bilingualism. Significance was based on the chi-square test, $P < 0.05$, as documented in Bernal Stoopen, "Binational collaboration in endangered species recovery efforts."

23. C. C. Chester, *Conservation across Borders: Biodiversity in an Interdependent World* (Washington, D.C.: Island Press, 2006).

24. Bernal Stoopen, "Binational collaboration in endangered species recovery efforts."

25. See ibid. Based in part on S. G. Clark, *Ensuring Greater Yellowstone's Future: Choices for Leaders and Citizens* (New Haven, Conn.: Yale University Press, 2008), and Chester, *Conservation across Borders*, and in line with the "generative" organizational model described in J. R. Gordon, *A Diagnostic Approach to Organizational Behavior* (Boston: Allyn and Baco, 1983), and R. Westrum, "An organizational perspective: Designing recovery teams from the inside out," in *Endangered Species Recovery: Finding the Lessons, Improving the Process*, edited by T. W. Clark, R. P. Reading, and A. L. Clarke, 327–50 (Washington, D.C.: Island Press, 1994).

26. Bernal Stoopen, "Binational collaboration in endangered species recovery efforts."

27. Gordon, *A Diagnostic Approach to Organizational Behavior*; Westrum, "An organizational perspective." R. W. Brislin and T. Yoshida elaborate on cross-cultural issues. See R. W. Brislin and T. Yoshida, eds., *Improving Intercultural Interactions: Modules for Cross-Cultural Training Programs* (Beverly Hills, Calif.: Sage, 1994).

28. In *Conservation across Borders*, Chester has elaborated on these ideas with respect to conservation across boundaries between countries and states. In *Ensuring Greater Yellowstone's Future*, Clark develops these themes in more detail regarding boundaries between lands under jurisdiction of separate state and federal agencies. Additional examples are summarized in R. P. Reading and B. Miller, eds., *Endangered Animals: Conflicting Issues* (Denver: Greenwood Press, 2000).

29. Bernal Stoopen, "Binational collaboration in endangered species recovery efforts."

Shared Ecosystem
Services

Shared Ecosystem Services

The authors in this section present a novel approach to transboundary conservation, framing it in terms of *shared ecosystem services*. Ecosystem services are the ways in which ecosystems and the species that make them up sustain and fulfill human life. The well-being of human society depends on the services of ecosystems, including the air we breathe, the water we drink, the food that nourishes us, and the aesthetic experiences that inspire our cultures and fulfill our lives.[1] In a transboundary context, drivers of environmental change in one country may affect the delivery and quality of ecosystem services and consequently the well-being of people in another country.[2] In such situations, ecosystem services can be used to frame transboundary conservation in the countries' mutual interest.

The four chapters in this section provide examples of ecosystem services shared by Mexico and the United States, demonstrating how drivers of change in one country can affect the delivery and quality of services in another country. The Millennium Ecosystem Assessment, an international effort to assess the status of the world's ecosystems, identified four ways by which ecosystems provide services. The chapters in this section demonstrate how the United States and Mexico share all four types of services: *supporting services*, or processes such as nutrient cycling and soil formation that are necessary to support biodiversity; *provisioning services*, or material benefits to humans, such as water or food; *regulating services*, or processes such as pollination, flood, and disease control that regulate other ecological processes; and *cultural services*, or those aspects of nature that provide humans with recreational, spiritual, or religious experiences.[3]

As with the conservation of shared species addressed in the previous section, the geographical scope of the ecosystem processes and services under discussion in this section ranges from the local to the continental. For instance, the services analyzed in the section's chapters include those provided by prairie dogs within grasslands in the immediate vicinity of the border and by the monarch butterfly, which has a continentwide range.

In chapter 11, Rodrigo Medellín explains how bats provide both regulating services—including pollination of and pest control for key crops in the United States and Mexico—and supporting services as seed dispersers. In chapter 12, Gerardo Ceballos and his coauthors stress the role of prairie dogs in maintaining the ecosystem processes that provide critical services in grasslands of the U.S.–Mexico borderlands. In chapter 9, Laura López-Hoffman, Robert Varady, and Patricia Balvanera present monarch butterflies as an example of a cultural service enjoyed by people in Mexico and the United States. In these examples, the mutual interest of both countries in protecting ecosystem services aligns well—efforts in either country to conserve migratory bats and butterflies will enhance services in both nations.

In contrast, the chapters also provide examples of situations where U.S. and Mexican interests in the services provided by water do not align well; efforts to increase water provisioning in one country can alter ecosystems in the other country, in turn impairing human well-being. For example, in chapter 10, Luis Calderon-Aguilera and Karl Flessa tie the reduction in Colorado River water flows to the upper Gulf of California to declining fisheries in Mexico, pointing out that the United States benefits from most of the ecosystem services provided by the Colorado River, whereas Mexico bears the majority of ecosystem costs. López-Hoffman, Varady, and Balvanera suggest that in these situations, innovative approaches such as new institutions and legal structures or transboundary payments for ecosystem services may be used to find common ground.

Notes

1. G. C. Daily, *Nature's Services: Societal Dependence on Natural Ecosystems* (Washington, D.C.: Island Press, 1997); Millenium Ecosystem Assessment, *Ecosystems and Human Well-Being: A Framework for Assessment* (Washington, D.C.: Island Press, 2003).

2. L. López-Hoffman, K. W. Flessa, and P. Balvanera, "Ecosystem services across borders: A framework for transboundary conservation policy," *Frontiers in Ecology and the Environment* (2009), doi:10.1890/070216.

3. Millennium Ecosystem Assessment, *Ecosystems and Human Well-Being*.

Chapter 9

Finding Mutual Interest in Shared Ecosystem Services

New Approaches to Transboundary Conservation

Laura López-Hoffman, Robert G. Varady, and Patricia Balvanera

In a Nutshell

- The United States and Mexico share provisioning ecosystem services (such as water), regulating services (such as crop pollination by bats), and the cultural services of migratory species (such as monarch butterflies).

- Actions and policies in one country may affect the delivery and quality of ecosystem services in another country, impacting human well-being.

- It is in the mutual interest of the United States and Mexico to conserve the ecosystems and ecosystem processes that provide their shared services.

- Strategies to protect specific services include: a treaty to distribute shared groundwater equitably; a treaty to protect the pollination services of migratory bats; an initiative to identify drivers of monarch butterfly decline and to prioritize areas for habitat conservation throughout North America.

- In addition, a broad mechanism, such as the Transboundary Environmental Impact Assessment, is needed to protect against transboundary impacts of actions in one country on ecosystem services of another country.

Introduction

When neighboring countries share ecosystems and the species that range across or fly over their borders, they share the services provided by those ecosystems and species as well.[1] The well-being of human society depends on the services of ecosystems, including the air we breathe, the water we drink, the food that nourishes us, and the aesthetic experiences that inspire our cultures and fulfill our lives.[2] Because the United States and Mexico share services, management actions and policies in one country can affect the delivery and quality of ecosystem services as well as people's well-being in the other country. In this essay, we suggest new approaches for conserving and protecting transboundary ecosystem services shared by Mexico and the United States.

In a previous essay, we advocated using the concept of ecosystem services as an organizing principle for transboundary conservation because it meets a central criterion of successful transboundary environmental policy—that discussions be framed in terms of mutual interests between countries, not in terms of rights and needs.[3] Actions taken in the mutual interests of two nations create incentives to work together rather than against one another. In our previous essay, we pointed out that the notion of interest—importance to human well-being and society—is inherent in the ecosystem services concept. We argued that if transboundary conservation were framed as the conservation of shared ecosystem services, the discussion would be transformed into one organized around mutual interests between countries.[4]

In this essay, we provide specific suggestions for how the ecosystem service concept can be used to frame U.S.–Mexico transboundary conservation in terms of mutual interests. We begin by presenting three cases that exemplify important services shared by the United States and Mexico. Next, we reflect on the lessons for developing transboundary policy that emerge from the three examples. We then suggest strategies for protecting the specific types of ecosystem services presented in the case studies, including new treaties and agreements. We conclude by suggesting ways of integrating the ecosystem service concept more generally into an existing binational environmental institution. Our discussion uses the Millennium Ecosystem Assessment (MA) conceptual framework of the relationship between ecosystem services and human well-being.

The MA is an international effort to assess the links between ecosystems, the services they provide, and human well-being. It identifies two types of drivers of ecosystem change: indirect drivers (social transformation such as population and economic growth) and direct drivers (management practices).[5] We have adapted this framework to elucidate how drivers in one country can affect the delivery of ecosystem services and human welfare in the other country (or in both countries) and how stakeholders might collaborate across international borders to protect shared ecosystem services (fig. 9.1).

The MA categorizes four ways in which ecosystems provide services. Material benefits to humans, such as water or food, are *provisioning services*. Processes such as pollination, flood, and disease control that

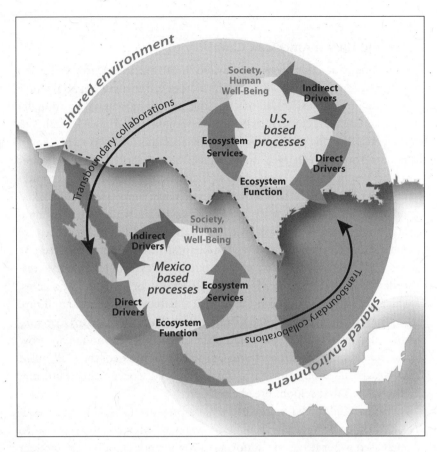

Figure 9.1. The Millennium Ecosystem Assessment (2003) framework as applied to transboundary conservation: drivers in one country can affect ecosystem services in the other country, and stakeholders can collaborate across international borders. Graphic design by Renee La Roi.

regulate other ecological processes are *regulating services*. Processes such as nutrient cycling and soil formation that are necessary to support biodiversity are *supporting services*. Those aspects of nature that provide humans with recreational, spiritual, or religious experiences are *cultural services*. We present three cases of ecosystem services shared by Mexico and the United States: (1) the provisioning services of transboundary groundwater by the All-American Canal in California; (2) the regulating services of long-nosed bats in pollinating agave crops; and (3) the cultural services of monarch butterflies.

Cases

Water and the All-American Canal

The All-American Canal was constructed in the 1940s to carry Colorado River water, a provisioning service, to farmers in California's Imperial Valley and people in San Diego. Millions of cubic meters of water seep annually from the unlined dirt canal, filtering into the aquifer under the Mexicali Valley in Mexico. The high-quality leaked water accounts for 10 to 12 percent of the aquifer's annual recharge, enhancing its water quality.[6] The seeped water is an inadvertent addition to Mexico's official allotment from the Colorado River under the 1944 International Water Treaty. Since 1942, the leaked water has been a source of irrigation and drinking water—provisioning services—for the residents of the Mexicali Valley. In addition, water seepage from the canal has created new wetland habitats that support biodiversity (fig. 9.2). The seepage has also created 6,000 hectares (around 15,000 acres) of wetlands on the Andrade Mesa (3,500 hectares, or around 8,600 acres, of which are in Mexico) that provide critical habitat to protected and rare bird species including the Yuma Clapper Rail *(Rallus longirostris yumanensis)*, endangered in the United States; the Large-billed Savannah Sparrow *(Passerculus sandwishensis rostratus)*, protected in Mexico; the Gull-billed Tern *(Gelochelidon nilotica)*, a Species of Special Concern in California; and three birds that seldom breed elsewhere.[7]

For years, California water users have pressured the U.S. Bureau of Reclamation to reduce the seepage of water to Mexico. In response to the increased water needs of California's growing population—the indirect driver—the bureau has decided to stop the transboundary leakage of water to Mexico. In mid-2007, it began to line sections of the canal with cement in order to prevent 83.5 million cubic meters of seepage annually. Direct drivers of change in the United States (i.e., new water-management strategies) will alter water deliveries in both countries: deliveries will increase in the United States but decrease in Mexico.

In 2005, a group of Mexican business and civic leaders and two California-based environmental nongovernmental organizations (NGOs) sued the Bureau of Reclamation in U.S. District Court, asserting that the canal lining would make the Mexicali aquifer "completely unusable" for the 1.3 million people of the Mexicali Valley, hurt the local economy, and destroy wildlife and wetlands in Mexico.[8] The lawsuit noted that although the bureau's 1994 and 2006 environmental impact assessments considered the lining's

Figure 9.2. The All-American Canal. *Inset:* A Landsat image showing seepage from the canal and the Andrade Mesa wetlands in Mexico. Photograph by Karl W. Flessa.

potential effect on wetlands in Mexico, they suggested wetland mitigation only for the United States.

The lawsuit was dismissed in July 2006. The court declared that the U.S. Constitution's Fifth Amendment protections against deprivation of property without due process do not apply to people outside U.S. territory and that disputes over international water treaties should be settled only through international diplomacy.[9] A 2006 waiver from the U.S. Congress prevented the court from considering whether the loss of wetland habitat in the Mexican portion of the Andrade Mesa would constitute violations of the U.S. Endangered Species Act, the National Environmental Policy Act, and the Migratory Bird Treaty.

There are currently no mechanisms for resolving groundwater conflicts between the United States and Mexico. The 1944 treaty distributed the surface waters of the Colorado and Tijuana rivers and the Rio Grande/ Río Bravo but did not address groundwater. In 1973, Minute 242 was added to the treaty, committing the countries to developing a mechanism for resolving groundwater disputes, but no real progress has been made.[10]

In fact, the Mexican section of the International Boundary and Water Commission/Comisión Internacional de Límites y Aguas—the binational body mandated to distribute surface waters—formally opposed the lining of the All-American Canal but was powerless to intervene because the canal is wholly inside the United States.

The difficulty of managing transboundary groundwater may in part be due to the nature of the service: efforts to increase water availability on one side of the border necessarily dictate a decrease on the other side, making it difficult but not impossible to find solutions to sharing groundwater that are of mutual interest and considered equitable by all parties. In addition, the "invisible" nature of water underground[11] may have contributed to the lack of binational groundwater management.

The Pollination Connection between Long-nosed Bats and Tequila

Bats provide the regulating service of pollinating agave, which is critical to tequila production in Mexico. Two species of long-nosed bats (genus *Leptonycteris*) are the principal pollinators of the blue agave plant *(Agave tequilana)*;[12] agave hearts are cooked and distilled to make tequila. Because sugar content is higher if the plant is prevented from flowering, Mexican corporate producers clone agave rather than allowing natural pollination, flowering, and reproduction. As a result, most large plantations in central Mexico consist of only one or two genetic varieties.[13] The consequences of low genetic diversity have been severe. In the late 1980s and again in 1996 and 1997, the genetically homogenous agave crops were devastated by pathogens, resulting in sizeable economic losses.[14] If corporate producers relied on natural bat pollination, the resulting genetically diverse agave crops would be less susceptible to diseases.[15]

The ecosystem services provided by bats are clearly critical to tequila production. However, the pollinator's future is uncertain; long-nosed bats are listed as endangered in both the United States and Mexico. Their habitat is threatened in both countries, particularly in the overwintering caves of the U.S.–Mexico borderlands (see chapter 11 in this volume). On both sides of the border, millions of bats have been burned, dynamited, or barred from their roosts by ranchers and cattlemen who mistake them for vampire bats. Bat caves have also been destroyed by urban development and highway construction.[16]

Corporate and small-scale tequila producers in Mexico are pursuing different types of interventions. Corporate producers, aware of the importance of genetically diverse agave crops, are using genetic engineering to increase diversity rather than relying on natural pollination processes.[17] Small, artisanal producers who depend on bat pollination and use many genetic varieties of *A. tequilana* as well as other agave species[18] are collaborating with conservation biologists to protect bats.[19]

The destruction of long-nosed bat habitat in both countries—resulting in declining bat populations and reduced pollination services—is caused by several indirect drivers: increased urban development and highway construction due to population growth in the United States and Mexico, as well as cultural fears that lead people to confuse pollinating bats such as the long-nosed bats with vampire bats.

The Monarch Butterfly and Aesthetic Fulfillment

People from Canada to Mexico experience wonder and a sense of aesthetic fulfillment when they witness the extraordinary migration of the monarch butterfly *(Danaus plexippus)*. Every fall, more than one hundred million monarch butterflies migrate from Canada and the United States to ten small mountaintops in the central Mexican states of México and Michoacán. The monarchs arriving at the oyamel fir *(Abies religiosa)* forests are several generations removed from those that left the previous year. Yet year after year they return to roost in the same area. The monarchs' aesthetic services in turn support the economic activity of ecotourism. The spectacular sight of a forest laden with butterflies is drawing ecotourists to the Monarch Sanctuary, boosting the economy of the village of Angangueo, Michoacán (fig. 9.3).

Monarch butterflies are in jeopardy throughout their range, however. In Mexico, illegal logging of oyamel firs is threatening the butterflies' winter ground.[20] In the monarchs' U.S. and Canadian summer grounds, the pollen of corn transgenically engineered to express insecticidal proteins may be harming them.[21] In Canada, milkweed *(Asclepias* spp.), the monarch's primary summer host plant and food source, is considered a noxious weed designated for eradication.[22] In the United States, intensive agricultural practices have reduced native vegetation around fields, resulting in loss of the milkweed plants the monarchs depend on for their fall migration to Mexico.[23]

Most significant butterfly-conservation efforts have until recently focused on the monarch's overwintering sites in Mexico. In 1986, the Mexican government proclaimed the oyamel forest sites a Biosphere Reserve and off-limits to logging.[24] International NGOs such as the World Wide Fund for Nature (or World Wildlife Fund, WWF) have been paying local people

Figure 9.3. Monarch butterflies. Photograph by Maria Isabel Ramírez Ramírez.

to abstain from logging in the oyamel forests.[25] Nonetheless, rates of defor-estation seem to be increasing in the reserve.[26] Only recently has significant attention been paid to preventing drivers of monarch decline in the U.S. and Canadian summer grounds. In April 2008, the trilateral Commission for Environmental Cooperation (CEC) and WWF-Mexico began an effort to identify drivers of monarch decline and to prioritize areas for habitat conservation throughout the butterflies' entire North American range.[27]

In sum, the drivers of the decline in monarch butterfly numbers are poverty in central Mexico (resulting in illegal logging), agricultural reliance on insecticidal genetically modified corn in the United States, and agricul-tural practices and weed-control policies resulting in a loss of milkweed in Canada. Although drivers of change are occurring in all three countries, the most significant conservation interventions have until recently focused only on Mexico.

Discussion

Several ideas that emerge from the case studies should be considered when framing conservation strategies to protect shared ecosystem services.

Transboundary ecosystem services extend beyond the border region. Under the 1983 La Paz Agreement,[28] 100 kilometers (62 miles) on either side of the political line was deemed the zone of concern for transboundary environmental protection. This 200-kilometer-wide (124-mile) strip was later adopted by the North American Free Trade Agreement (NAFTA) and by its side provision for border environmental infrastructure investment, the North American Agreement on Environmental Cooperation (NAAEC) (see chapters 1 and 17 in this volume). Although the geographic scope of water provided by the All-American Canal falls within the 200-kilometer strip, the services provided by bats and butterflies clearly demonstrate how the ecological connections between the United States and Mexico range deep into each country (fig. 9.4). Furthermore, the case studies show how actions in the interior of one country can affect the ecosystem services received by people in the heart of another country. Binational efforts to protect transboundary ecosystem services must recognize that the scope of ecosystem services extends beyond the border region.

Many types of ecosystem services are shared by the two countries. Although transboundary discussions about water resources have focused on the importance of water as a provisioning service for human welfare, the concept of ecosystem services has not been used to frame the conversation. To date, most policy and academic discussions about shared species have focused on the species per se and not on the importance of species populations and ecosystem processes for supporting human well-being. It is important to recognize that the species shared by the United States and Mexico impart critical supporting, regulating, and cultural services.

Drivers of change are similar. Although we present different types of ecosystem services, common drivers affect them. In the All-American Canal case, increased water demands from southern California's growing population will decrease water supply in the Mexicali Valley. In the bat and butterfly examples, urbanization and land-use change degrade critical habitat supporting the service. In all three examples, it is notable that the effects of expanding populations and land- and water-use intensification are felt by stakeholders who are far removed geographically and even beyond international boundaries.

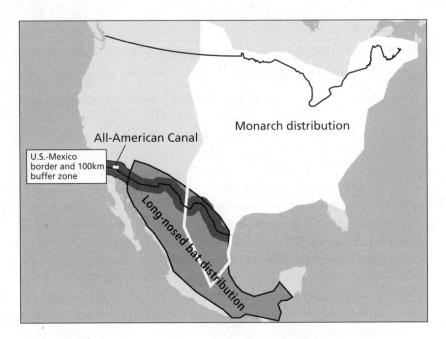

Figure 9.4. Monarch butterfly and long-nosed bat distributions in North America. The political line between the United States and Mexico is in black; the 100-kilometer (62-mile) fringe is shaded dark gray; and the All-American Canal is shown as a white bar.

Diverse stakeholders must be included in binational decision making to protect shared ecosystem services. In the long-nosed bat case study, corporate producers and small-scale tequila producers in Mexico value bat ecosystem services very differently; small-scale producers are aiming to protect bats and their services, whereas corporate stakeholders are seeking to replace pollination services with technological solutions for increasing genetic diversity. Binational efforts to protect shared services must clearly include diverse stakeholders.

Local stakeholders are particularly important to developing binational collaborations to protect ecosystem services. In all three examples, local stakeholders initiated the transboundary efforts to protect shared services; governments became involved only after civil society groups, including environmental NGOs and researchers, laid the groundwork. In the All-American Canal situation, it was local Mexican stakeholders, not the Mexican federal government, who sued the U.S. government over loss of ecosystem services.

In fact, the local stakeholders felt that their needs were not being adequately addressed by either their state government or their federal government.[29] Strategies to protect shared ecosystem services should facilitate transboundary collaboration between stakeholders at all scales, but it is particularly important to provide local stakeholders with means to work cooperatively across the border.

Policy Options for Protecting Shared Ecosystem Services

In this section, we suggest ways of using the ecosystem service concept to frame transboundary conservation as being in the mutual interest of both countries. First, we outline strategies to protect the types of ecosystem services presented earlier, discussing bats and monarch butterflies, then address the more complicated case of water provisioning. We conclude by advocating a broad mechanism to protect against unintended impacts of actions and policies in one country on ecosystem services in another country; we suggest that the CEC revive the Transboundary Environmental Impact Assessment (TEIA) and adapt it to encompass ecosystem services.

Regulating services provided by bats

Because bats provide ecosystem services in both countries, efforts to protect bat populations in one country should also bolster the services they provide in the other country. An effort is currently under way to establish the North American Bat Conservation Alliance—similar to the North American Bird Conservation Initiative (see chapter 7 in this volume). This alliance would be wise to monitor the transboundary ecosystem services provided by migratory bats and document their value. A treaty to protect migratory bats, similar to the Migratory Bird Treaty, might eventually be developed; the importance of bat transboundary services can be used to demonstrate that the United States and Mexico have a mutual interest in developing such a treaty.

Cultural services of monarch butterflies

The challenge in the case of monarch butterflies is not in aligning national interests—that is, any effort to protect monarchs in one country will benefit stakeholders in all countries—but in helping each country accept its responsibility in protecting the monarch. Until recently, most conservation attention focused on the Mexican government's failure to halt logging in the monarch's winter grounds.[30] Nonetheless, drivers of change in monarch populations also originate in the United States and Canada. To protect monarchs and their cultural services, Canada should consider removing milkweed from the federal list of noxious weeds. In the United States, the agricultural practices that reduce native vegetation and milkweed around fields must be modified. The CEC and WWF-Mexico's recently launched North American monarch-conservation program should emphasize working with the United States and Canada to change these practices.

Provisioning services of shared groundwater

In the All-American Canal situation, the interests of the United States and Mexico do not naturally align because efforts to increase water services on one side of the border necessitate a decrease in those services on the other. In situations where binational interests do not align well, innovative approaches such as new institutions, legal structures, and creative trans-boundary collaborations may be used to find common ground.

First, the United States and Mexico would be wise to honor their commitment under Minute 242 of the 1944 Water Treaty and develop a mechanism for equitably resolving groundwater disputes along the border. The adoption of the Bellagio Draft Treaty on Transboundary Groundwater, developed in 1987 by experienced groundwater managers and scientists from around the world, would provide such a mechanism. Although the draft treaty specifically contemplated the U.S.–Mexico border, its general principles were designed to apply to transboundary groundwater situations worldwide. Its main premise is that the twin needs of utilizing and conserving shared groundwater should be balanced between countries on "an equitable and reasonable basis." The draft treaty provides a framework for the sharing of transboundary groundwater but does not create specific obligations between the countries.[31]

As a possible step toward establishing a cross-border groundwater treaty, the U.S. Congress enacted the U.S.–Mexico Transboundary Aquifer Assessment Act in 2006. This act promotes cooperation between U.S. and Mexican border states wherein appropriate entities in the two countries will be involved in "conducting a hydrogeologic characterization, mapping, and modeling program for priority transboundary aquifers."[32]

Second, even in the absence of a groundwater treaty, creative solutions to both surface and groundwater conflicts might involve transboundary payments for ecosystem services. For example, in the All-American Canal situation, U.S. and Mexican stakeholders concerned about the loss of biodiversity in Mexico's Andrade Mesa wetlands might buy existing Colorado River water rights in Mexico and dedicate the water to wetland protection. Mexico's national water law was recently amended to allow for "environmental use"; U.S. water law notably does not allow for such uses. Income from selling existing water rights might partially offset Mexican agricultural losses due to groundwater reductions. Two NGOs, the U.S.–based Sonoran Institute and Mexico's Pronatura Noroeste, are using a similar approach to secure water for restoring the Colorado River delta (see chapter 2 in this volume).

The CEC and TEIA

In addition to the aforementioned ways of conserving specific types of ecosystem services, a broad mechanism is needed to protect against the unintended impacts of actions and policies in one country on ecosystem services in another country.

The CEC, as the institutional arm of the NAAEC, is the ideal platform for such a mechanism. In chapter 16 of this volume, Steve Mumme and his coauthors describe what they call the CEC's SOS functions—*scaling* ecological linkages across North America, *organizing* networks of stakeholders, and *spotlighting* issues of transboundary environmental concern through investigations. The CEC's trinational mandate is broad enough to encompass, for example, the continental reach of the ecosystem services provided by long-nosed bats or monarch butterflies. In addition, its broad topical mandate (see the areas of concern in chapter 16) would allow it to examine issues such as ecosystem services that lie at the nexus of economic development and the environment. Finally, the CEC has existing mechanisms for incorporating a wide variety of stakeholders, including local people, into transboundary conservation.

Not only do the CEC's broad geographic and topical mandate and history of working with diverse stakeholders make it ideal for promoting the conservation of shared ecosystem services, but it has also already developed an instrument that can encompass transboundary impacts on services. In the NAAEC, the newly formed CEC was charged with developing a trilateral mechanism for assessing the potential transboundary impacts of actions and policies in one country on the environment of another country. With an expert group from the United States, Mexico, and Canada, the CEC developed TEIA. Under the proposed TEIA, any project requiring an environmental impact assessment would also require an evaluation of possible transboundary environmental impacts. If a project or action were judged to have transboundary impacts, representatives from the impacted country would be included in the originating country's planning process.

TEIA would not oblige a country to halt a project that might negatively impact the people or environment of another country; its objective is simply to provide decision makers with timely information on possible transboundary environmental consequences.

Were TEIA to encompass ecosystem services, the Canadian government would have to consider the transboundary impacts of its noxious weed policies on butterfly populations in the United States and Mexico. In the All-American Canal example, under TEIA, the U.S. Bureau of Reclamation would have had to consider the transboundary impacts of lining the canal on the wetlands in Mexico before it made its decision. The existence of a TEIA mechanism might not have stopped the bureau from lining the canal, but it might have persuaded the bureau to fund wetland restoration and mitigation in Mexico.

Unfortunately, despite NAAEC's commitment to implement TEIA, the United States, Canada, and Mexico have not adopted this mechanism. In the mid-1990s, when TEIA was first proposed, the U.S. Department of State and its Mexican counterpart were not supportive of the idea. We are hopeful, however, that the renewed spirit of binational cooperation between the Barack Obama and Felipe Calderón administrations, coupled with the former's recognition of the importance of upholding international agreements, might move the United States and Mexico to honor their commitment to implement TEIA.

Conclusions

It is in the mutual interest of Mexico and the United States to develop programs to sustain and protect shared ecosystem services. They should honor their commitment in Minute 242 to the 1944 Water Treaty and develop a mechanism similar to the Bellagio Draft Treaty that would provide for the equitable sharing of transboundary aquifers. They should consider developing a treaty for the conservation of migratory bats that recognizes that bats' ecosystem services are important to both agriculture and human well-being. Efforts to protect continentwide cultural symbols such as the monarch butterfly must recognize that the drivers of change affecting the butterfly population are occurring in both countries. The CEC would be wise to examine transboundary ecosystem services as one of their areas of concern. However, much more is needed: the TEIA mechanism should be expanded to consider transboundary impacts on ecosystem system services, and the United States and Mexico would be well served to honor their commitment in NAFTA by adopting this mechanism.

Mexico and the United States share many ecosystem services, including many not covered in this chapter. The water flowing across the U.S.–Mexico border through rivers and aquifers provides critical supporting and provisioning services in both countries; migratory species such as bats and butterflies travel many hundreds of miles over the border, imparting critical regulating services and cultural services in the heart of each country. Shared ecosystem services link the well-being of people in both nations because actions in one country may affect the services received by people in the other country. It is in the mutual interest of the United States and Mexico to work together to protect the ecosystems and ecosystem services that support the well-being of their citizens.

Notes

1. L. López-Hoffman, R. Varady, K. Flessa, and P. Balvanera, "Ecosystem services across borders: A framework for transboundary conservation policy," *Frontiers in Ecology and the Environment* (2009), doi:10.1890/070216.

2. Millennium Ecosystem Assessment, *Ecosystems and Human Well-Being: Our Human Planet* (Washington, D.C.: Island Press, 2005).

3. López-Hoffman et al., "Ecosystem services across borders." See also L. Susskind, W. Moomaw, and K. Gallagher, eds., *Transboundary Environmental Negotiation: New Approaches to Global Cooperation* (San Francisco: Jossey-Bass, 2002), and A. T. Wolf, "Shared waters: Conflict and cooperation," *Annual Review of the Environment and Resources* 32 (2007): 241–69.

4. López-Hoffman et al., "Ecosystem services across borders."

5. Millennium Ecosystem Assessment, *Ecosystems and Human Well-Being: A Framework for Assessment* (Washington, D.C.: Island Press, 2003).

6. U.S. Bureau of Reclamation, *All-American Canal Lining Project: Final Environmental Impact Statement/Environmental Impact Report* (Washington, D.C.: U.S. Bureau of Reclamation, 1994).

7. O. Hinojosa-Huerta, P. L. Nagler, Y. Carrillo-Guerrero, E. Zamora-Hernández, J. García-Hernández, F. Zamora-Arroyo, K. Gillon, and E. P. Glenn, "Andrade Mesa wetlands of the All-American Canal," *Natural Resources* 42 (2002): 899–914.

8. "District court dismisses seven of eight claims challenging the All-American Canal lining project," *California Water Reporter* (May 2006), available at http://www.argentco.com/htm/f20060506.722351.htm, accessed September 24, 2009.

9. Ibid.

10. S. P. Mumme, "Minute 242 and beyond: Challenges and opportunities for managing transboundary groundwater on the Mexico–U.S. border," *Natural Resources Journal* 40 (2000): 341–79.

11. H. Ingram, "Transboundary groundwater on the U.S.–Mexico border: Is the glass half full, half empty, or even on the table?" *Natural Resources Journal* 40 (2000): 185–89.

12. L. E. Eguiarte and H. T. Arita, "The natural history of tequila and mezcal, or the tequila connection, take two," *Tropinet* 18 (2007): 1–11.

13. R. Dalton, "Alcohol and science: Saving the agave," *Nature* 438 (2005): 1070–71.

14. A. G. Valenzuela Zapata and G. P. Nabhan, *¡Tequila! A Natural and Cultural History* (Tucson: University of Arizona Press, 2003).

15. S. Arizaga, E. Ezcurra, E. Peters, F. R. Arellano, and F. Vega, "Pollination ecology of *Agave macroacantha* (Agavaceae) in a Mexican tropical desert. I. Floral biology and pollination mechanisms," *American Journal of Botany* 87 (2002): 1004–10.

16. S. Walker, "Mexico–U.S. partnership makes gains for migratory bats," *BATS* 13, no. 3 (1995), available at http://www.batcon.org/index.php/media-and-info/bats-archives.html?task=viewArticle&magArticleID=717, accessed September 25, 2009.

17. Dalton, "Alcohol and science."

18. P. Colunga-García Marín and D. Zizumbo-Villarela, "Tequila and other agave spirits from west-central Mexico: Current germplasm diversity, conservation, and origin," *Biodiversity Conservation* 16(6) (2006): 1653–67.

19. R. Medellín, Instituto de Ecología, Universidad Nacional Autónoma de México, personal communication, May 30, 2008.

20. C. Galindo-Leal, *Perdida y deterioro de los bosques en la Reserva de la Biosfera Mariposa Monarca 2005–2006* (Mexico City: World Wildlife Fund–Mexico, 2006).

21. L. H. Jesse and J. J. Obrycki, "Field deposition of Bt transgenic corn pollen: Lethal effects on the monarch butterfly," *Oecologia* 125, no. 2 (2000): 241–48.

22. Monarch Watch, *Status, Distribution, and Potential Impacts of Noxious Weed Legislation* (May 22, 2007), available at http://www.monarchwatch.org/read/articles/canweed2.htm, accessed September 24, 2009.

23. L. P. Brower, L. S. Fink, and P. Walford, "Fueling the fall migration of the monarch butterfly," *Integrative and Comparative Biology* 46(6) (2006): 1123–42.

24. *Diario oficial: Organo del gobierno constitucional de los Estados Unidos Mexicanos* (Mexico City: n.p., 1986), 33–41.

25. M. Missrie and K. Nelson, *Direct Payments for Conservation: Lessons from the Monarch Butterfly Conservation Fund* (Twin Cities: College of Natural Resources, University of Minnesota, 2005).

26. Galindo-Leal, *Perdida y deterioro de los bosques.*

27. Commission for Environmental Cooperation (CEC), "Commission on Environmental Cooperation and World Wildlife Fund–Mexico seek input on monarch conservation" (2008), available at http://www.cec.org/news/, accessed May 1, 2008.

28. "Agreement Between the United States of America and the United Mexican States on Cooperation for the Protection and Improvement of the Environment in the Border Area" (La Paz Agreement), signed on August 14, 1983, full text available at http://www.epa.gov/usmexicoborder/docs/LaPazAgreement.pdf/.

29. A. Heras, "Apatía federal ante revestimiento del Canal Todo Americano: Labriegos de BC," *La Jornada* (Mexico City), May 8, 2007.

30. Galindo-Leal, *Perdida y deterioro de los bosques;* Missrie and Nelson, *Direct Payments for Conservation.*

31. R. D. Hayton and A. E. Utton, "Transboundary groundwaters: The Bellagio Draft Treaty," *Natural Resources Journal* 29 (1989): 663–721.

32. House Bill 469, Senate Bill 214, 109th Cong., 2d sess., May 2006.

Add Water?

Transboundary Colorado River Flow and Ecosystem Services
in the Upper Gulf of California

Luis E. Calderon-Aguilera and Karl W. Flessa

In a Nutshell ────────────────────────────────

- The United States benefits from the ecosystem services provided by the Colorado River and its delta, whereas Mexico bears the majority of ecosystem costs associated with the river's use and development.

- A reduction of freshwater to the Colorado River delta degrades habitat for imperiled and economically important species.

- Unsustainable fishing and shrimp-harvesting practices negatively affect long-term economic prosperity while also directly endangering the rarest cetacean in the world, the vaquita porpoise.

- Both countries would benefit economically from restoring certain ecosystem services to the Colorado delta and upper Gulf of California.

- As little as 25 cubic meters (33 cubic yards) per second freshwater delivery to the delta paired with ecosystem-based management of fishing systems would benefit commercial and sport fisheries, shrimp production, and tourism, and would protect a number of endangered species.

Introduction

In most years, the Colorado River no longer reaches the sea. The upstream diversion of nearly all Colorado River water in the United States and Mexico has altered the hydrology, geology, biology, and ecology of the river's estuary—the mixing zone between river water and seawater—in the upper Gulf of California. The changes have degraded the environment and reduced the value of the ecosystem services once supported by the river's flow (see chapter 9 in this volume). Populations of shellfish have been reduced, nursery grounds for commercially valuable shrimp and finfish have been degraded, and an endangered species of fish is at risk because of both fishing practices and the lack of river water. The degradation of the estuary and upper gulf is the result of water-allocation policies in the United States and Mexico and of fisheries practices in Mexico. The United States enjoys most of the economic benefits, whereas Mexico bears most of the environmental costs.

Environmental restoration of the estuary and upper gulf will yield economic and social benefits to both countries. Restoration of flows of Colorado River water and a shift to ecosystem-based management of Mexican fisheries are needed. Existing binational frameworks might allow the United States and Mexico to equitably allocate portions of their Colorado River water for purposes of restoration.

The science is clear; now, political action is needed. This chapter asks: Can these two nations, within the scope of their systems of water allocation, devise policies that will promote the restoration of the estuary and the ecosystem services that the estuary provides? Based on historical flow data, field sampling, and data gathered from a number of sources, we show that adding a small amount of freshwater to the upper Gulf of California would make a big difference in terms of ecological and economic benefits.

Background

In the United States, major estuaries such as the Sacramento River/San Francisco Bay, Chesapeake Bay, and the Everglades/Florida Bay system are the focus of massive restoration efforts. These estuaries have been affected by the pollution of their incoming rivers, diversions of their incoming rivers' flow, coastal development, and overfishing.[1] Efforts at restoration have been motivated both by U.S. law (e.g., the Clean Water Act, the Endangered Species Act) and by the realization that estuaries provide valuable ecosystem services to society. In effect, the public has demanded these efforts.

Estuaries provide many ecosystem services, the goods and services provided to human society by nature—for example, habitat for the spawning, development, and subsistence of commercially important fisheries; habitat for migratory and resident birds; recreation; pollutant filtration; and shoreline protection.[2] Although it is both difficult and controversial to assign a dollar value to estuaries, Robert Costanza and his colleagues estimate that tidal marshes and mangroves provide ecosystems services worth U.S.$22,832 per hectare per year.[3] That figure is overly precise, but it is the largest estimated value among the twelve biomes surveyed by these researchers. Estuaries clearly provide valuable benefits to human society.

The Colorado River estuary presents a special transboundary case. Which nation should bear responsibility for the decline in ecosystem services resulting from the diversion of the river's water? The United States diverts approximately 90 percent of the river's flow before it reaches the Mexican border. The U.S. economy benefits greatly, but at the cost of environmental damage in Mexico. Once in Mexico, the remaining 10 percent of the flow is diverted to provide support for Mexican agriculture and cities. The ecosystem services provided by a functioning estuary would largely benefit Mexico (discussed more fully later). Would the United States also benefit? Can these two nations, with different economies and different legal systems for allocating water, work together to restore the estuary for mutual benefit? We think so and outline our recommendations later in this chapter.

A River Diverted: Impact on the Colorado River Estuary and Upper Gulf of California

The Colorado River is born in the Rocky Mountains of Colorado. Along its way to the Gulf of California, its water is used by seven U.S. states (Wyoming, Colorado, Utah, New Mexico, Arizona, Nevada, and California) and in Mexico by Baja California and Sonora. The Colorado River delta, as described by Godfrey Sykes in 1937, encompassed 8,611 square kilometers (861,100 hectares) extending over parts of Arizona, California, Baja California, and Sonora.[4] The river once delivered its entire annual discharge of approximately 14 million acre-feet (about 17.3 billion cubic meters) of freshwater to the upper Gulf of California, creating an enormous estuary that extended from the river's mouth 70 kilometers (44 miles) south to San Felipe, Mexico (see chapter 2 in this volume).[5] Today, the trickle of water that sometimes reaches the estuary is salty runoff from farmers' fields and effluent from water-treatment plants.

The U.S. Bureau of Reclamation's regulation and diversion of Colorado River flow following the completion of upstream dams has had profound effects on the upper Gulf of California and Colorado River delta ecosystems:

Altered circulation of the upper Gulf of California. Prior to upstream diversions, lower-density freshwater from the river flowed southward at the surface of the upper Gulf of California, with higher-salinity seawater flowing along the bottom, back toward the river's mouth. Today, high-salinity water, formed by the evaporation of seawater in the confines of the river's mouth, sinks and flows southward, and seawater flows toward the mouth at the surface.[6] Oceanographers refer to such circulation as "antiestuarine."

Altered coastline. Prior to upstream diversions, the river delivered approximately 147 million tons of sediment per year to its delta.[7] Coarser sediments were deposited near the mouth of the river, and the finer-grained silt and mud were carried seaward. Most sediment today is trapped in reservoirs behind the upstream dams. After nearly one hundred years of river regulation, the delta is slowly retreating toward the mainland.

Reduced abundance of shellfish. Prior to dams and diversions, population densities of shellfish ranged from twenty-five to fifty specimens per square meter.[8] In contrast, surveys today show densities from two to seventeen specimens per square meter, a reduction of as much as 94 percent from predam values. This sharp decline has been attributed to the decrease in nutrients once supplied by the river.[9]

Altered species composition of shellfish. The Colorado delta clam *(Mulinia coloradoensis)*, once the most abundant bivalve mollusk inhabiting the intertidal and shallow subtidal zone of the delta, is now rare.[10] Geochemical, isotopic markers in this clam's shell indicate that it grew in low-salinity water—a mixture of river water and seawater that is no longer present. This species' decline might have resulted from the reduction in salinity or the reduction in nutrients caused by the upstream diversion of Colorado River water.[11]

Changes in the estuary's food web. As the Colorado delta clam's population declined, its predators either switched to other prey or declined in abundance. The Colorado delta clam was an important source of food for predatory crabs and snails.[12] Predation by crabs and snails is identified by the distinct marks on the shells of predam *Mulinia coloradoensis.* Other predators, such as fish or birds, that were either higher in the food web or that did not leave traces of their behavior are also likely to have been affected.

Reduction in growth rate of an endangered fish. Totoaba *(Totoaba macdonaldi)* was once commercially exploited and prized as a sport fish in the upper gulf. The species spawns and spends at least the first year of juvenile development in the Colorado River estuary.[13] Otoliths (earbones) from live-caught and predam specimens show that this species grew more quickly and reached sexual maturity sooner in the era before upstream diversions.[14] Along with overfishing, this change in life history contributed to the species' decline.

In addition to these before-and-after comparisons of the estuary's species, other evidence points to the importance of the river's flow to the estuary's ecosystem:

River flow increases populations of larval shrimp. Since the early 1960s, river water has rarely reached the estuary and the upper gulf. However, abnormally wet years and full reservoirs in the 1980s and 1990s caused U.S. water managers to release water to Mexico in excess of the treaty obligation. Because Mexico has no capacity to store water from such "controlled releases," the river flowed all the way to the upper Gulf of California during brief periods. Populations of postlarval shrimp in the estuary increased following the controlled releases.[15]

Shrimp catches increase following increases in river flow. Shrimp catches in the towns of El Golfo de Santa Clara and Puerto Peñasco in Sonora and San Felipe in Baja California were higher one year after controlled releases of Colorado River water.[16] This finding, along with the linkage between river water and larval shrimp, indicates that the Colorado River plays a major role in supplying nutrients to the delta's marine life as well as in increasing the suitable nursery habitat for shrimp.[17]

Increased flow benefits a commercially important fish. Gulf corvina *(Cynoscion othonopterus)* is a valuable fishery in the upper Gulf of California. This species returns every year to the mouth of the river to spawn, where it is intensively fished using gill nets.[18] During years of excess Colorado River flow (1993 and 1999), gulf corvina inhabited and grew in the estuary's less-saline waters.[19] It had been absent from catches for thirty years, but it rebounded in 1997, four years after the excess flow of 1993. Between 1958 and 1993, there had been little freshwater input to the upper Gulf of California. These data suggest that an influx of river water creates a larger nursery ground for these fish and in turn increases their adult population in subsequent years.

In addition to this previous work, evidence presented here demonstrates the importance of river water to the biology of the region's commercially important fish. The thirteen species (except Pacific hake, *Merluccius productus*) that constitute 98 percent of the commercial catches here require brackish (low-salinity) water during their early development (table 10.1). Figure 10.1 shows river flow to the gulf and catches in the same year.

Table 10.1. Relative landings of the main commercial species from the upper Gulf of California.

Scientific name	English common name[a]	Trophic Level[b]	Catch contribution to overall landings during study period (%)	Cross-correlation coefficient (r)	Lag (yrs)
Cynoscion othonopterus	gulf corvina	4.0	31	0.31	−4
Micropogonias megalops	bigeye croaker	3.2	25	0.36	0
Litopenaeus spp.	shrimp	3.0	11	0.23	−1
Scomberomorus sierra	Pacific sierra	4.5	7	0.22	−1
Mustelus sp.	shark	4.1	6	0.31	0
Merluccius productus	Pacific hake	3.9	5	0.16	−6
Dasyatis brevis	whiptail stingray	3.5	3	0.03	−3
Squatina californica	Pacific angel shark	4.14	3	0.12	−2
Rhinobatos productus	guitar fish	3.5	2	0.15	−4
Epinephelus acanthistius	gulf coney	3.2	2	0.42	0
Paralichthys aestuarius	Cortez flounder	3.6	1	0.32	−4
Callinectes	crab	3.3	1	0.27	−4
Mugil cephalus	mullet	2.0	1	0.40	0

[a] Common names from J. S. Nelson, E. J. Crossman, H. Espinosa-Pérez, L. T. Findley, C. R. Gilbert, R. N. Lea, and J. D. Williams, *Common and Scientific Names of Fishes from the United States, Canada, and Mexico,* Special Publication no. 29 (Bethesda, Md.: American Fisheries Society, 2004).

[b] Trophic level, ecology, and growth rate data are from R. Froese and D. Pauly, eds., *FishBase* (2005), World Wide Web electronic publication, available at http://www.fishbase.org, accessed July 2007. (Some from the closest taxon when the species is not available.)

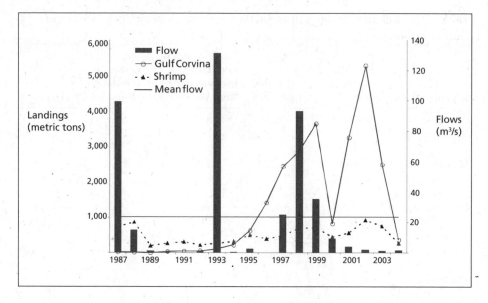

Figure 10.1. Catches (in metric tons) of principal commercial species from the upper Gulf of California and flows (in cubic meters per second [m³/s]) of the Colorado River, for the period from 1987 to 2004.

An analysis of freshwater flow and commercial landings suggests that the size of catches is correlated with river flow. Data on freshwater discharges of the Colorado River at the international boundary for the period from 1987 to 2004 were obtained from the International Boundary and Water Commission/Comisión Internacional de Límites y Aguas (IBWC/CILA), and commercial landings statistics were obtained from official records (see fig. 10.1).[20] From 1987 to 2004, there were five years (1987, 1993, 1997, 1998, and 1999) in which river flow was above average (24 cubic meters [31 cubic yards] per second).[21] The results of the analysis strongly suggest a relationship between Colorado River flow and fisheries catches in the upper Gulf of California.

The results also suggest that the traditional single-species approach to fisheries management by the Comisión Nacional de Acuacultura y Pesca (CONAPESCA), the Mexican national fisheries agency, is inadequate. Although the single-species approach makes sense because fishing gear, quotas, regulations (minimum legal size, fishing season, etc.), and permits are oriented to catch a target species, it has proved to be insufficient to prevent overfishing. Ecosystem-based management may be more effective because its goal is not the health of a single species, but the long-term health of the ecosystem—an ecological community together with its environment. The goal of the latter approach is to maintain ecosystem health and sustainability. As demonstrated in this chapter, the influx of Colorado River water is vital to the health of the upper Gulf of California's ecosystem and needs to be considered in the management of fisheries in the region.

Plenty of Problems to Go Around: Overfishing in the Upper Gulf of California

Not all of the degradation of the Colorado River estuary can be blamed on the decrease in the supply of water from the Colorado River. Damage from fishing gear, fishing practices, and overfishing is a serious problem. At present, Mexican fishing trawlers plow the soft bottoms and destroy the three-dimensional structure of the ocean bottom, reducing habitat complexity and biodiversity. Maintaining the ecosystem's structure is fundamental for the health of invertebrate fisheries. Illegal fishing is rampant by both local subsistence fishers and commercial entities. Accurate accounting of catches is needed for any management strategy. Shrimp is the most economically valuable resource in the upper Gulf of California, but at current rates of exploitation, shrimp catches not only are unsustainable but are not even profitable to Mexican fishers.[22] The current fishing season goes from mid-September to mid-May, but catches from February to May are less than 2 percent of the total.[23] In addition, bycatch, or the incidental capture and killing of nontarget species, is very high in the shrimp fishery. A shortening of the fishing season by CONAPESCA would rebuild the population, protect spawning stock and nontarget species, and increase yields.[24]

To determine the health of a fishing population, researchers utilize the mean trophic level of the fisheries catch. The mean trophic level represents a proxy variable related to the feeding habits of the species in catches; a high mean level indicates that the fishery is sustained by higher trophic levels of carnivorous species—a healthy fishery—whereas a low mean level may indicate overfishing (i.e., "fishing down the food web"). In the upper Gulf of California, the mean trophic level of the catch ranged from 3.39 in 1988 to 3.81 in 2002 (fig. 10.2). This increase might suggest that the ecosystem was becoming healthier, but it was actually due to the impressive resurgence of a single species, the gulf corvina, a species that was virtually absent from the area until the brief resumption of Colorado River flow in the 1980s and 1990s. If the gulf corvina catches are removed from the analysis, the mean

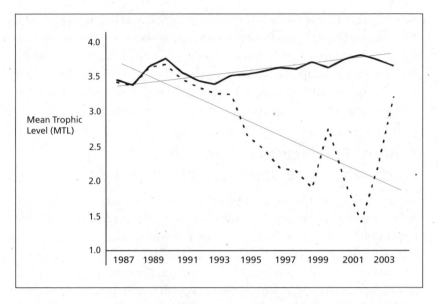

Figure 10.2. Mean trophic level (MTL) of landings in the upper Gulf of California *(solid line)* and without landings of corvina *(dashed line)*. Note that the MTL shows a sharp decline since 1987 if landings of gulf corvina are not included.

trophic levels decrease (fig. 10.2). These results demonstrate both the impact of unsustainable fishing practices and the low freshwater inputs.

Fishing practices also threaten the survival of the vaquita *(Phocoena sinus)*. This endemic species of harbor porpoise now has the unfortunate distinction of being the most endangered cetacean in the world. Fewer than two hundred individuals remain, and the species is at risk of extinction because individuals get caught and drown in the gill nets commonly used in the upper gulf.[25] There is no evidence suggesting that the decline in Colorado River flow has endangered this species.

Economic and Social Impacts of Environmental Degradation in the Study Area

In spite of fisheries decline and environmental degradation in the upper gulf, fishing is still a very important economic activity, providing more than 7,500 direct jobs and at least twice that number of indirect jobs.[26] The value of catches in the upper gulf averages U.S.$7 million annually (see fig. 10.3). In recent years, resorts, marinas, and other tourist developments in

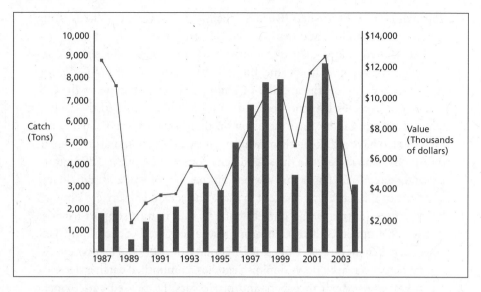

Figure 10.3. Volume (in tons, *bars*) and value of catches (in thousands of dollars, *line*) of the principal species in the upper Gulf of California. Landings statistics are from Comisión Nacional de Acuacultura y Pesca (CONAPESCA), Secretaría de Agricultura, Ganadería, Desarrollo Rural, Pesca y Alimentación, *Anuarios estadísticos de pesca* (2004), at http://www.sagarpa.gob.mx/conapesca, accessed April 2005. Prices in pesos were converted to U.S. dollars at the mean annual exchange rate. They are prices from fisher to first buyer; market prices can be higher.

Puerto Peñasco and San Felipe have generated jobs, although they tend to be temporary construction or low-wage domestic service jobs.

Shrimp is the most economically important fishery in the Gulf of California.[27] Two species, *Litopenaeus stylirostris* (blue shrimp) and *Farfantepenaeus californiensis* (brown shrimp), use the upper gulf as a nursery ground.[28] Catches of shrimp in San Felipe have decreased from more than 500 tons per season on average during the 1980s to less than 300 tons at present. Both the lack of river water flowing from upstream and unsustainable fishing practices can be blamed.

Because fishing is currently the primary economic activity in the region, increased water to the upper gulf would mean more than just better catches.[29] Increased water would have direct social and economic impacts. Catches in 2002 were worth more than U.S.$12 million (fig. 10.3) and represent direct and indirect jobs for the region's communities. In Puerto Peñasco, there are more than 3,150 fishers, as well as 4,100 people dependent on them;

approximately 2,850 fishers live in San Felipe, and 1,500 in Golfo de Santa Clara.[30] Their families and many businesses depend on them.

The United States also experiences environmental and social consequences of reduced flows and poor fishing practices in the Colorado River estuary and the upper Gulf of California. Two species are on the U.S. endangered species list: totoaba and vaquita. Their value to the United States is recognized by their inclusion on that list. Both the United States and Mexico have a stake in their recovery. In addition, both nations benefit directly from an increase in the region's environmental health: U.S. ecotourists and recreational fishers are attracted to healthy ecosystems. Employees of Mexican tourist services benefit in turn. The ecosystem services supplied by river flow (nutrients and nursery habitat) that in turn support fisheries in Mexico's upper Gulf of California also affect the U.S. economy. U.S. consumers benefit directly from a nearby source of seafood. Furthermore, as shown here, decreased flows diminish catches. Diminished catches decrease employment of Mexican fishers and processors. Decreased employment drives immigration—legal and illegal—to the United States. As what is sometimes called the First Law of Ecology states, "Everything is connected to everything else."

Solutions and Policy Alternatives

As in so many cases of environmental degradation, in this case there is not a single cause, and there is not a single villain. There is plenty of blame to go around. Mexico must internally come to grips with the problems caused by overfishing in the upper Gulf of California. But this necessity does not excuse the United States from solving the lack of river flow to the estuary, a problem it has caused, and the associated negative environmental and socioeconomic impacts. The causes of these problems have now been scientifically demonstrated. The United States and Mexico share responsibility and must jointly participate in effecting solutions. The joint solutions are the focus of our suggestions for improved transboundary policy.

The simplest solution, of course, is for the United States and Mexico to allocate sufficient water for the Colorado River to reach the sea once again. As we suggest here, a flow of as little as 25 cubic meters (33 cubic yards) per second would have beneficial effects, but more study is needed to determine if this flow would be sufficient for a substantive restoration. How can the goal of restoration be achieved?

The IBWC/CILA—the binational agency responsible for administering the U.S.–Mexico border and the treaties regulating water allocations between the two countries—should be expanded. A binational scientific advisory committee is needed to determine the water needs of the delta's riparian, wetland, and estuarine environments. IBWC/CILA might serve as a forum to negotiate the equitable share of water from each country's existing allocation. The formation, if not the funding, of such a group is already possible under Minute 306 of the 1944 International Water Treaty between the United States and Mexico. This so-called environmental minute (in effect, an amendment to the treaty) calls for studies to determine the delta's water needs and to formulate approaches to allocate supplies. Such a group must have sufficient financial resources to accomplish its goals.

The Commission for Environmental Cooperation, which sprang from the North American Agreement on Environmental Cooperation (the environmental side agreement to the North American Free Trade Agreement), should address the impact of binational economic development on the allocation and use of Colorado River water. It should facilitate the development of strategies for restoring the river's estuary. It is critical that the commission confront the reality that environmental costs to Mexico are arising from economically beneficial use of the river's water in the United States. Addressing this issue is a matter of simple fairness, but would benefit both countries (see chapter 16 in this volume).

Mexico's fisheries agency, CONAPESCA, should transition from single-species fisheries management to ecosystem-based management. It should also engage in efforts to restore the river flows that would benefit the upper Gulf of California fisheries.

Ongoing efforts by environmental groups to establish a water trust—a supply of water taken from allocations for agriculture, agricultural return flow, and treated effluent—that might also be used for riparian, wetland, and estuarine restoration in Mexico should be supported (see chapter 2 in this volume).

Conclusion: From Science to Policy

Scientific work during the past decade has documented that the lack of Colorado River flow and unsustainable fishing practices have caused environmental degradation of the Colorado River estuary and upper Gulf of California.

The concept of transboundary ecosystem services provides a framework to identify the benefits and costs of transboundary allocation policies regarding Colorado River water.[31] The river once connected snowfields of the Rocky Mountains, deep in the United States, with nursery grounds of shrimp and fish in Mexico's Gulf of California, supporting ecosystem services in Mexico. The development of irrigated agriculture and the growth of human populations drove changes in flows of water, now diverting most of it before it reaches the border with Mexico and all of it before it reaches the sea. The central issue here is not the border itself. It is the disparity in the two countries' economic and political power.

Both unilateral action and bilateral action are needed. Mexican fisheries management must shift to a sustainable, ecosystem-based scheme. The United States must acknowledge that water management within its borders has consequences for ecosystem services beyond its borders. Restoration of ecosystem services in the Colorado River estuary and upper Gulf of California is critical, but it is a binational challenge that can be solved through negotiations facilitated by an existing binational agency (IBWC/CILA) and within the framework of existing international treaties. Both countries will benefit from the solutions. The problem is clear, the path to these solutions is wide open; what is missing is the will to proceed.

Acknowledgments

This work was partially funded by University of California Institute for Mexico and the United States, Consejo Nacional de Ciencia y Tecnología (CR-01-08,) and Centro de Investigación Científica y de Educación Superior de Ensenada (622143) grants to Luis Calderon-Aguilera and by U.S. National Science Foundation grants to Karl Flessa. Comments from editors and anonymous reviewers significantly improved this paper. David Garfield, Emily McGovern, and Laura López-Hoffman helped with reviewing and editing. Victor Moreno and several students participated in the fieldwork.

Notes

1. H. K. Lotze, H. S. Lenihan, B. J. Bourque, R. H. Bradbury, R. G. Cooke, M. C. Kay, S. M. Kidwell, M. X. Kirby, C. H. Peterson, and J. B. C. Jackson, "Depletion, degradation, and recovery potential of estuaries and coastal seas," *Science* 312 (2006): 1806–9.

2. G. C. Daily, ed., *Nature's Services: Societal Dependence on Natural Ecosystems* (Washington, D.C.: Island Press, 1997); Millennium Ecosystem Assessment, *Millennium Ecosystem Assessment Reports* (Washington, D.C.: Island Press, 2005), available at http://www.millenniumassessment.org/en/Index.aspx.

3. R. Costanza, R. d'Arge, R. de Groot, S. Farber, M. Grasso, B. Hannon, K. Limburg, S. Naeem, R. V. O'Neill, J. Paruelo, R. G. Raskin, P. Sutton, and M. van den Belt, "The value of the world's ecosystem services and natural capital," *Nature* 387 (1997): 253–60. See also G. Heal, *Nature and the Marketplace: Capturing the Value of Ecosystem Services* (Washington, D.C.: Island Press, 2000), and National Research Council, *Valuing Ecosystem Services* (Washington, D.C.: National Academies Press, 2005).

4. G. Sykes, *The Colorado Delta*, Publication no. 460 (Washington, D.C.: Carnegie Institution, 1937).

5. N. Carbajal, A. Souza, and R. Durazo, "A numerical study of the ex-ROFI of the Colorado River," *Journal of Marine Systems* 12 (1997): 17–31; C. A. Rodriguez, K. W. Flessa, M. Téllez-Duarte, D. Dettman, and G. Avila-Serrano, "Macrofaunal and isotopic estimates of the former extent of the Colorado River estuary, upper Gulf of California, Mexico," *Journal of Arid Environments* 49 (2001): 185–95.

6. M. F. Lavín and S. Sánchez, "On how the Colorado river affected the hydrography of the upper Gulf of California," *Continental Shelf Research* 19 (1999): 1545–60.

7. R. W. Thompson, *Tidal Flat Sedimentation on the Colorado River Delta, Northwestern Gulf of California*, Geological Society of America Memoir no. 107 (Boulder, Colo.: Geological Society of America, 1968).

8. M. Kowalewski, G. E. Avila-Serrano, K. W. Flessa, and G. A. Goodfriend, "Dead delta's former productivity: Two trillion shells at the mouth of the Colorado River," *Geology* 28 (2000): 1059–62.

9. G. E. Ávila-Serrano, K. W. Flessa, M. A. Téllez-Duarte, and C. E. Cintra-Buenrostro, "Distribution of the intertidal macrofauna of the Colorado River Delta, northern Gulf of California, Mexico," *Ciencias Marinas* 32(4) (2006): 649–61.

10. C. A. Rodriguez, K. W. Flessa, and D. L. Dettman, "Effects of upstream diversion of Colorado River water on the estuarine bivalve mollusk *Mulinia coloradoensis*," *Conservation Biology* 15 (2001): 249–58.

11. C. E. Cintra-Buenrostro, K. W. Flessa, and G. E. Avila-Serrano, "Who cares about a vanishing clam? Trophic importance of *Mulinia coloradoensis* inferred from predatory damage," *Palaios* 20 (2005): 296–302.

12. Ibid.

13. C. A. Flanagan and J. R. Hendrickson, "Observations on the commercial fishery and reproductive biology of the totoaba, *Cynoscion macdonaldi*, in the northern Gulf of California," *Fisheries Bulletin* 74 (1976): 531–44; M. A. Cisneros-Mata, G. Montemayor-López, and M. J. Román-Rodríguez, "Life history and conservation of *Totoaba macdonaldi*," *Conservation Biology* 9 (1995): 806–14.

14. K. Rowell, K. W. Flessa, D. L. Dettman, M. J. Román, L. R. Gerber, and L. T. Findley, "Diverting the Colorado River leads to a dramatic life history shift in an endangered marine fish," *Biological Conservation* 141 (2008): 1138–48.

15. E. A. Aragón-Noriega and L. E. Calderon-Aguilera, "Does damming the Colorado River affect the nursery area of blue shrimp *Litopenaeus stylirostris*

(Decapoda:Penaeidae) in the upper Gulf of California?" *International Journal of Tropical Biology Conservation* 48(4) (2000): 867–71.

16. M. S. Galindo-Bect, E. P. Glenn, H. M. Page, K. F. L. Galindo-Bect, J. M. Hernandez-Ayon, R. L. Petty, J. García-Hernández, and D. Moore, "Penaeid shrimp landings in the upper Gulf of California in relation to Colorado River freshwater discharge," *Fishery Bulletin* 98 (2000): 222–25.

17. Aragón-Noriega and Calderon-Aguilera, "Does damming the Colorado River affect the nursery area of blue shrimp?"; L. E. Calderon-Aguilera, E. A. Aragón-Noriega , H. A. Licón, G. Castillo-Moreno, and A. Maciel-Gómez, "Abundance and composition of penaeid postlarvae in the upper Gulf of California," in *Contribution to the Study of East Pacific Crustaceans*, vol. 1, edited by M. E. Hendrickx, 281–92 (Mexico City: Anales del Instituto Ciencias del Mar y Limnología, Universidad Nacional Autónoma de México, 2002).

18. L. E. Calderon-Aguilera, J. M. García Caudillo, and A. Díaz de León, "Pesquerías: Abundancia que requiere manejo," in *Región Golfo de California: Síntesis sobre su sociedad, economía y recursos naturales,* edited by E. Bolado, 25–30 (Guaymas, Mexico: Conservación Internacional, Región Golfo de California, 2006).

19. K. Rowell, K. W. Flessa, D. L. Dettman, and M. Román, "The importance of Colorado River flow to nursery habitats of the Gulf corvine *(Cynoscion othonopterus),*" *Canadian Journal of Fisheries and Aquatic Sciences* 62 (2005): 2874–85.

20. International Boundary and Water Commission, "Daily flow of the Colorado River at the Southerly International Boundary" (2004), available at http://www.ibwc.state.gov/wad/DDQSIBCO.htm), accessed July 2005; Secretaría de Medio Ambiente, Recursos Naturales y Pesca (SEMARNAP), *Anuarios estadísticos de pesca* (Mexico City: SEMARNAP, 2000); Comisión Nacional de Acuacultura y Pesca (CONAPESCA), Secretaría de Agricultura, Ganadería, Desarrollo Rural, Pesca y Alimentación, *Anuarios estadísticos de pesca* (2004), available at http://www.sagarpa.gob.mx/conapesca, accessed April 2005.

21. The cross-correlation coefficient r represents the degree to which the catch is correlated with a previous year's flow. A lag of -2 means that the catch of that particular species is correlated with the flow from two years prior to that catch. The lag represents the time needed for the juvenile fish to grow to their minimum catch size. The higher the r value, the more the variation in catch is explained by the variation in river flow. The theoretical maximum value of r is 1, indicating a perfect correlation between catch and the previous year's river flow. Values in table 10.1 range from 0.03 to 0.42. Although none of these r values are statistically indistinguishable from 0.0, the small number of years in the analysis and the fact that all the values are positive are strongly suggestive of a relationship between Colorado River flow and fisheries catch in the upper Gulf of California.

22. Calderon-Aguilera, García Caudillo, and Díaz de León, "Pesquerías."

23. Instituto Nacional de la Pesca, "Carta nacional pesquera 2000," *Diario Oficial de la Federación*, August 17 and 28, 2000.

24. Calderon-Aguilera, García Caudillo, and Díaz de León, "Pesquerías."

25. L. Rojas-Bracho, R. R. Reeves, and A. Jaramillo-Legoretta, "Conservation of the vaquita *Phocoena sinus*," *Mammal Review* 36 (2006): 179–216.

26. Calderon-Aguilera, García Caudillo, and Díaz de León, "Pesquerías."

27. Ibid.

28. Calderon-Aguilera et al., "Abundance and composition of penaeid post-larvae"; E. A. Aragón-Noriega and L. E. Calderon-Aguilera, "Age and growth of shrimp postlarvae in the upper Gulf of California," *Aqua Journal of Ichthyology and Aquatic Biology* 4 (2001): 99–104.

29. Calderon-Aguilera, García Caudillo, and Díaz de León, "Pesquerías"; R. Cudney-Bueno and P. Turk Boyer, *Pescando entre mareas del alto Golfo de California* (Puerto Peñasco, Mexico: CEDO, 1998).

30. J. L. Gallardo García, "Recursos pesqueros en el alto Golfo de California: La influencia de las señales del mercado," master's thesis, El Colegio de la Frontera Norte, Tijuana, Mexico, 1996; F. J. De la Cruz, "Políticas de manejo y aspectos socioeconómicos en la Reserva de la Biosfera y Delta del Río Colorado: El caso de la pesca ribereña en San Felipe, B.C.," master's thesis, Centro de Investigación Científica y de Educación Superior de Ensenada, El Colegio de la Frontera Norte, Baja California, 2002.

31. L. López-Hoffman, R. G. Varady, K. W. Flessa, and P. Balvanera, "Transboundary ecosystem services: Policy options for services shared by Mexico and the United States, *Frontiers in Ecology and the Environment* (2009), doi:10.1890/070216.

...aining Transboundary Ecosystem ...rvices Provided by Bats

Rodrigo A. Medellín

In a Nutshell

- Bats provide essential ecosystem services that support human well-being, including pollination, seed dispersal, and pest control.
- Thirty-four species of bats provide ecosystem services to people in the U.S.–Mexico borderlands, including support of cotton, corn, and agave production.
- One-third of these thirty-four species are included in national or international lists of threatened and endangered species due to habitat destruction.
- Ecosystem services, including those provided by bats, are not accounted for in long-term management and policy decisions in the United States and Mexico.
- Multilateral conservation efforts in North America are beginning to recognize bat conservation as a major priority for human well-being and ecosystem maintenance.
- The North American Bat Conservation Alliance, along with other programs, must increase educational, research, and conservation actions within the transboundary region.

Introduction

Bats are the second most diverse order of mammals; in terms of number of species, they surpass nearly all other groups of terrestrial vertebrates. Their diversity is not only taxonomical; no other group of mammals provides a more diverse array of ecosystem services, including seed dispersal, pollination, and insect pest control. The transboundary region shared by Mexico and the United States is important habitat for several bat species that play an important role in these ecosystem services. Many millions of bats spend a large proportion of their lives in the transboundary region, most significantly during the summer, when millions of bats of several species arrive from central and southern Mexico. Other species remain in the region year-round, hibernating in caves, mines, and crevices during the winter, and emerging from these roosts in the spring. The species of bats that utilize habitats in the transboundary region provide critically important ecosystem services to people in the United States and Mexico, and all throughout their range.

Bats are one of the most rapidly declining groups of mammals (order Chiroptera) in the world. In the U.S.–Mexico transboundary area, where they represent a significant number of all species present, they face severe threats that have already caused significant population losses. It is critical that bat ecosystem services are considered in national accounting for the environment in both countries, including all productive activities linked to them, most prominently agriculture. Because natural ecosystem function-ing relies heavily on bat-mediated ecological processes, society on both sides of the international border should take up conservation of bats as a top priority.

In this chapter, I describe in detail the ecosystem services provided by bats and the importance of these services to human well-being. I also identify ways to evaluate bat ecosystem services economically. Finally, I discuss con-servation measures necessary on both sides of the international boundary and argue that reaching across the border is imperative in order to ensure that these services continue into the future.

Background

Biological diversity is increasingly recognized as providing services vitally important for the life-support systems of the world.[1] Animals, plants, and other living organisms determine the balance of atmospheric gases and are responsible for the ecological functioning of all ecosystems in the world. In an attempt to demonstrate and document the extraordinary importance of biological diversity, many researchers have begun to invest time and effort in documenting these *ecosystem services*.[2] A simple online search of scientific publications reveals 1,330 articles published on the subject of ecosystem services in the past thirty years, accelerating heavily in the past fifteen years. More and more ecosystem services are identified each year, and the concept is increasingly integrated into national and international decision-making processes. However, ecosystem services are not yet included in any national accounting systems, at least in part due to the extreme complexities of measuring services that have previously been taken for granted. Some of the first ecosystem services documented as supporting human well-being include those provided by insects, coastal ecosystems, soil animals, mangroves, medicinal plants, and wildlife in general.[3] An important set of ecosystem services are those provided by bats.

There are more than 1,100 species of bats in the world.[4] This remarkable taxonomic diversity is mirrored in the extreme variety of life histories they exhibit. From solitary species to extremely gregarious species, and from those that roost in caves to those that roost in hollow trees or vegetation clumps, bat biology is a clear reflection of taxonomic diversity. Bats follow the familiar biogeographic pattern of becoming more diverse as one approaches the tropics (although some bat species even range into the polar circles). Yet bats are also very diverse and abundant in the Mexico–U.S. transboundary area (fig. 11.1). In fact, some of the largest concentrations of mammals in the world occur precisely on either side of the Mexico–U.S. border. From Texas and Tamaulipas in the east to Arizona, California, Sonora, and Baja California in the west, this region contains numerous caves, some of which harbor huge colonies of bats, ranging up to several million individuals. Such large colonies are rare outside the Mexico–U.S. borderlands, so conserving the colonies in this area and the ecosystems that sustain them is paramount to maintaining bat ecosystem services in the transboundary region and beyond.

At least 34 of Mexico's 138 bat species are found in the transboundary region (see table 11.1).[5] All but 3 of these 34 species are insectivores. The remaining 3 species are pollinivorous species responsible for the sexual reproduction of many species of plants in the transboundary region. At least 13 and as many as 20 of the 34 bat species in the region migrate between Mexico and the United States, and possibly farther. A few species, such as the hoary bat *(Lasiurus cinereus)* may migrate from Mexico to Canada.

Of the 34 bat species, 16 gather singly or in groups of tens of bats, 9 form colonies of hundreds of individuals, and the other 9 gather in large colonies of thousands, hundreds of thousands, or up to millions of individuals. One of the latter, the Mexican free-tailed bat *(Tadarida brasiliensis mexicana)*, is representative of the 24 species that live in caves, further highlighting the need to protect the region's cave ecosystems from destruction.[6]

Bats are among the vertebrate groups experiencing the greatest population declines worldwide, which are potentially causing negative effects on the ecosystem services they provide.[7] At present, approximately one-third of the bat species documented in the transboundary area are listed as at-risk species

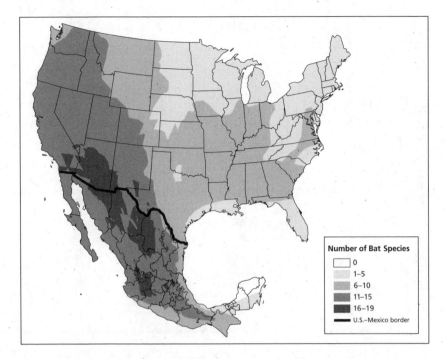

Figure 11.1. Species richness of bats occurring in the transboundary area between Mexico and the United States.

under the following listings: the U.S. Endangered Species Act (2 species), the Mexican List of Species at Risk of Extinction (6 species), and the "Red List" of the International Union for Conservation of Nature (7 species).

Ecosystem Services

The tremendous variation in bat lifestyles and ecologies is also largely responsible for bats' wide range of ecosystem services. The major ecosystem services provided by bats fall into three broad categories: seed dispersal, pollination, and pest control.

Seed Dispersal

Seed dispersal is a crucial ecosystem service; between 50 and 90 percent of tropical tree and shrub species depend on vertebrates to move their seeds. Seed dispersal also prevents competition between parent plants and their offspring, allows colonization of new or unoccupied habitats, maintains

Table 11.1. Bats inhabiting the transboundary area between Mexico and the U.S. and some biological traits.

Family and Species Name	Distribution	International Union for Conservation of Nature	Norma Oficial Mexicana 059, listing species at risk of extinction (NOM059)	U.S. Endangered Species Act	Do they live in caves?	Typical group size	Migratory?
—Mormoopidae—							
Mormoops megalophylla	U.S. to northern South America				Yes	Thousands	Yes
Choeronycteris mexicana	Southern U.S. to Guatemala	Lower Risk	Threatened	Endangered	Yes	1–10	Yes
—Phyllostomidae—							
Leptonycteris curasoae	U.S. to northern South America	Vulnerable	Threatened	Endangered	Yes	Thousands	Yes
Leptonycteris nivalis	U.S. to Guatemala	Endangered	Threatened	Endangered	Yes	Hundreds	Yes
Macrotus californicus	U.S. to northwestern Mexico	Vulnerable			Yes	Hundreds	

—Vespertilionidae—

Species	Distribution					
Antrozous pallidus	U.S. to central Mexico			Yes	Hundreds	Yes
Eptesicus fuscus	Canada to northern South America			Yes	Thousands	Yes
Euderma maculatum	U.S. to central Mexico	Special Protection	Endangered	Yes	1–10	?
Idionycteris phyllotis	U.S. to southern Mexico			Yes	Tens	?
Lasionycteris noctivagans	U.S. to northeastern Mexico	Special Protection			1–10	?
Lasiurus blossevillii	Canada to Chile and Argentina				1–10	Yes
Lasiurus borealis	Canada to northern Mexico				1–10	Yes
Lasiurus cinereus	U.S. to Chile and Argentina				1–10	Yes
Lasiurus ega	U.S. to Argentina				1–10	?
Lasiurus intermedius	U.S. to Central America				1–10	?
Lasiurus xanthinus	U.S. to central Mexico				1–10	?

Table 11.1. Continued

Family and Species Name	Distribution	International Union for Conservation of Nature	Norma Oficial Mexicana 059, listing species at risk of extinction (NOM059)	U.S. Endangered Species Act	Do they live in caves?	Typical group size	Migratory?
Myotis auriculus	U.S. to Guatemala				Yes	Hundreds	
Myotis californicus	Alaska to Guatemala				Yes	Thousands	
Myotis ciliolabrum	Canada to central Mexico				Yes	Thousands	
Myotis evotis	Canada to Baja California, Mexico		Special Protection		Yes	Hundreds	
Myotis lucifugus	Alaska to central Mexico				Yes	Thousands	
Myotis thysanodes	Canada to Mexico				Yes	Hundreds	
Myotis velifer	U.S. to Central America				Yes	Hundreds of thousands	
Myotis volans	Alaska to central Mexico				Yes	Hundreds	
Myotis yumanensis	Canada to central Mexico				Yes	Thousands	

Species	Range	Status	Population		
Nycticeius humeralis	Canada to north-eastern Mexico		Hundreds		
Pipistrellus hesperus	U.S. to central Mexico		1–10	Yes	?
Pipistrellus subflavus	U.S. to Central America		1–10	Yes	
Plecotus towsendii	U.S. to southern Mexico	Vulnerable	Hundreds	Yes	
—Molossidae—					
Eumops perotis	U.S. to Argentina		Tens		Yes
Eumops underwoodi	U.S. to Central America	Lower Risk	Tens		Yes
Nyctinomops femorosaccus	U.S. to Mexico		Tens	Yes	Yes
Nyctinomops macrotis	U.S. to South America		Tens	Yes	Yes
Tadarida brasiliensis	U.S. to Chile and Argentina	Lower Risk	Hundreds of thousands	Yes	Yes

Figure 11.2. Lesser long-nosed bat *(Leptonycteris curasoe)* at cardón cactus. Photograph by Marco Tschapka.

gene flow between otherwise isolated populations, and ensures succession and recovery of degraded forests, particularly in the tropics.

Bats are efficient seed dispersers in tropical and subtropical ecosystems, providing better and more copious dispersal than frugivorous (fruit-eating) birds. Bats disperse up to three times more seeds per square meter per 24-hour period than birds.[8] Although most research on seed dispersal by bats is from tropical ecosystems (especially rain forests), it is known that a few bat species provide this service in the Mexico–U.S. borderlands area: the lesser long-nosed bat *(Leptonycteris curasoe)* (fig. 11.2), the Mexican long-nosed bat *(L. nivalis)*, and the hog-nosed bat *(Choeronycteris mexicana)*. All are migratory species present in the transboundary area only during the summer months. The two *Leptonycteris* species are considered endangered in the United States and threatened in Mexico, and the hog-nosed bat is considered threatened in Mexico. The migratory nature of these three species makes them very good seed dispersers; they can fly up to 30 kilometers (19 miles) one-way from their roosts to their feeding grounds in a single night.[9]

These species' diet consists mostly of the fruit, nectar, and pollen of several dozen plant species, including the fruit of columnar cacti. Columnar cacti such as saguaro *(Carnegiea gigantea)*, cardón *(Pachycereus pringlei)*, and organ pipe *(Stenocereus thurberi)* have large, typically fleshy, juicy, and sweet fruits with hundreds of very small seeds. Each fruit has between 300 and 1,700 seeds or more, depending on the species and region.[10] Once the fruit is ripe, the spine-covered fruit cover breaks open, exposing the bright red to purple flesh, making it available to fruit-eating vertebrates, such as lizards, birds, bats, and humans. Many local people in central and northern Mexico harvest columnar cactus fruits for consumption, and in several areas these fruits are an important cash crop. They are harvested in the southwestern United States as well.

The geographic range of these three bat species spans from southern Mexico to extreme southwestern Texas, New Mexico, and Arizona. As a result, the seed-dispersal service provided by bats spans a large section of the transboundary area and extends far south into the deserts and the coniferous and tropical deciduous forests of central and southern Mexico.

Conservation of frugivorous bats is an important action that is still pending in many regions. Artificial roosts have been used to promote seed dispersal and forest regeneration by these species, although there is still some discussion about their effectiveness.[11]

Pollination

Pollination services by animals have been estimated at several to many billions of dollars annually. Bats pollinate hundreds of different plant species across the Americas, many of them critically important for both ecosystem functioning and human well-being. It has been estimated that bats pollinate the flowers of about 250 genera of vascular plants.[12] To put this information in perspective, there are 1,500 species of cultivated crop plants in the world, and birds pollinate about 3.5 to 5.4 percent of these plants, whereas bats pollinate about 6.8 to 10.7 percent.[13] Thus, the value of bat-mediated pollination is clearly an important support for food production as well as agricultural activities and markets, although specific valuations are problematic and are yet to be calculated.

At the northern end of the distribution of nectar-feeding bats, precisely within the transboundary area between Mexico and the United States, three endangered species of migratory, pollinating bats are critical for the reproduction of ecologically and economically important species of columnar cacti and a few dozen species of agaves called "century plants" (genus *Agave*). These bats carry out agave pollination, ensuring cross-breeding and enhancing agave genetic diversity, which are in turn key to minimizing the impacts of disease.[14] The three species of bats most associated with agave pollination—the long-nosed bats, *L. curasoe* and *L. nivalis*, and the hog-nosed bat, *Choeronycteris mexicana*—face extinction risks in the United States and Mexico.[15]

Agaves play an extraordinarily important role in maintaining the ecological function of deserts, xeric scrublands, and subtropical forests. They are also directly useful to humans in a number of ways. For instance, agaves can slow or stop soil degradation; they are being used on other continents to stabilize soil along roadsides. In addition, they are a source of natural fibers (sisal) that continue to be used in many countries around the world. Agaves also provide food to humans; whole panicles of the unopened flowers can be found for sale in markets across central Mexico. They also provide a cultural service because their fleshy, large leaves are used in central Mexico to wrap goat meat and cook it in a barbeque-style dish.

But the most well-known use of agaves is in the production of tequila and mescal. The sugar content of agaves is highest just before the plant blooms; an agave's potential for alcohol production is best at this moment.[16] As a result, many millions of plants across Mexico are prevented from blooming and are harvested just before the flowering stage. The tequila and mescal industries account for several hundred million dollars every year.

Two activities pose serious threats to the endangered pollinating bat species: habitat destruction, on the one hand, and roost destruction and vandalism, on the other.[17] Habitat destruction can be curbed by developing a management plan for protected areas throughout the transboundary region. Because of their economic value and diversity of uses, agaves will always be promoted by humans. Nonetheless, a cooperative management plan that allows a minimum number of agave to bloom so that bats can feed on them, continue their migration, and ensure agave genetic diversity is necessary. To mitigate roost destruction and vandalism, strong stakeholder education programs are necessary from big cities to small towns; such programs have been successful in other areas.[18] Collaboration both across sectors of society such as academia, government, and nongovernmental organizations and within the international arena is crucial.[19]

Pest Control

Insect pests are a major source of agricultural losses. Predator-prey interactions regulate insect pest populations, and the ecological effects of predators on the prey (pest) populations are crucial for agricultural practices. Understanding these interactions is crucial to sustain healthy levels of food production and to validate for the general public the goods and benefits provided by biodiversity.[20]

Many bat species use the Mexico–U.S. border region as summer grounds. One of these species is the Mexican free-tailed bat (*Tadarida brasiliensis mexicana*, a.k.a the Brazilian free-tailed bat). Many Mexican free-tailed bat colonies have suffered drastic declines due to vandalism and destruction of their roosting caves.[21] The pest-control service provided by the Mexican free-tailed bat is one of the very few ecosystem services for which researchers have actually assessed a dollar value. This bat is an insectivorous (insect-eating) species that preys primarily on moths, including pest species such as the Noctuids corn earworm moth *(Helicoverpa zea)*, the fall army worm *(Spodoptera frugiperda)*, and the cabbage looper *(Trichoplusia ni)*.[22]

Every summer, maternity colonies of Mexican free-tailed bats form in the transboundary area of northern Mexico and southern United States from Texas to Arizona; when the females migrate to the borderlands, most males remain farther south in central Mexico. The females travel to the northern end of the species distribution looking for plentiful food resources and a hot climate in which to give birth to their young. During the lactation process, they greatly increase their energetic needs—every night, each twelve-gram lactating female consumes six to eight grams of insects,[23] providing cotton growers in the United States and Mexico with an unsuspected and surprisingly helpful ecosystem service. A large bat colony can destroy many tons of insect pests every night.

Researchers have determined the value of bat pest control based on estimates of damage to cotton bolls prevented by bats and cost savings from the associated decrease in the need for insecticide applications. In the south-central part of Texas alone—known as the "winter garden region" with approximately 100 square kilometers (39 square miles) of cotton fields—the value of this bat pest regulation is on the order of U.S.$700,000 annually.[24] It has also recently been documented that bats provide pest control in Neotropical agroforestry and forested ecosystems.[25]

In addition to directly controlling populations of insect pests, these bats provide another related and valuable service: helping prevent the evolution of pesticide resistance. Genetically engineered crops have rapidly become established in technified agricultural fields, such as those in the Texas winter garden region. Cotton varieties enhanced with the toxic *Bacillus thuringiensis* (or "Bt") genes are becoming widespread across the Americas, especially in the United States. However, additional insecticide applications are still necessary to control insects, at least in Bt cotton. The combination of Bt cotton and supplemental insecticide applications kills up to 95 percent of the targeted pests.[26] The surviving 5 percent, although already affected by pesticides, can survive to reproduce if they are not killed by some other means—thus increasing the chances for these insect pests to become evolutionarily resistant to Bt.

Bat consumption of resistant insects diminishes the need for supplemental application of insecticides. It may ultimately delay the need for new insecticides to kill the few insects that survive Bt poison. According to Paula Federico and her colleagues, insectivorous bat predation of agricultural pests creates economic benefits by reducing the frequency of required spraying and by delaying the development of new pesticides. The agricultural economics of both Bt and conventional cotton production are more profitable whenever large numbers of insectivorous bats are present.[27]

Both Mexico and the United States are already actively working on the conservation and recovery of the caves that contain large colonies of insectivorous bats, but more work and more international coordination and cooperation are necessary along the transboundary region and beyond, given these species' migratory nature.[28]

Discussion and Conclusions

The ecosystem services provided by bats are very broad in scale and scope. Bat-mediated ecological processes benefit humans and ecosystems both near to and far from the Mexico–U.S. border, although the benefits have to date been studied in only a few areas. There is still much to be learned about the services provided by bats, but it is clear that bat conservation and recovery should be major priorities of cooperative conservation efforts in North America.

The conservation of biological diversity is on the agenda of virtually every biologist around the globe. A major part of the battle to conserve biodiversity lies in articulating the values of ecosystem services provided by biodiversity and functioning ecosystems for the benefit of human well-being.[29] Ecosystem services provided by specific taxa or groups of animals or plants are rarely, if ever, considered in national systems of economic accounting. Accounting of the gross domestic product of virtually every country in the world fails to recognize ecosystem services as assets or even as externalities. In this context, the case of bats is illustrative, demonstrating the importance of their services and how little the general public understands the very intimate links that connect us to bat biodiversity. It is crucial that we appreciate the ways that bats, though often unseen, provide benefits to us on a daily basis. We must direct our conservation activities to maintain habitats for bats.

The Association of Fish and Wildlife Agencies and the Trilateral Committee for Wildlife and Ecosystem Conservation and Management, a cooperative body established in 1996 among the three federal governments of Canada, Mexico, and the United States, are promoting the establishment and implementation of the North American Bat Conservation Alliance. The three federal governments are beginning to recognize the need for protecting bats across the continent. Although it has taken time, bats are on the way to becoming an example of transcontinental cooperation for wildlife and ecosystem conservation; the efforts to protect bats are similar to the efforts made by Partners in Flight and the North American Bird Conservation Initiative to protect migratory birds (see chapter 7 in this volume).

Transboundary conservation efforts on behalf of bats are the result of various stakeholders' and conservation programs' joined efforts to maximize bat protection. Groups such as the Western Bat Working Group, the South Eastern Bat Working Group, the North Eastern Bat Group, the Program for Conservation of Mexican Bats; entities of the Canadian, U.S., and Mexican federal governments; as well as Canadian and U.S. provincial and state governments have consolidated forces into a unified effort to establish objectives, strategies, and priorities.

The North American Bat Conservation Alliance is taking a three-pronged strategy of research, environmental education, and conservation actions. This strategy is similar to other successful programs such as the Mexico–U.S. Program for Conservation of Migratory Bats (PCMM), established in 1994. PCMM has expanded from initially working only on recovering three species of migratory bats to including all migratory, endemic, and species at risk in Mexico. Through cooperation among academic institutions, government agencies, and nongovernmental organizations, PCMM has either stabilized the decline or promoted the recovery of many species, including a supposedly extinct bat species;[30] it has also trained dozens of students and professionals in Mexico and other countries and promoted the creation of new programs in Bolivia, Costa Rica, Guatemala, and other countries. This program and its partners have now coalesced into the Latin American Network for Conservation of Bats (in Spanish, Red Latinoamericana para la Conservación de Murciélagos).[31] The PCMM is also pioneering the documentation of bat ecosystem services. Starting in the summer of 2008, PCMM personnel are recording the appearance of bulbils—tiny agave clones that appear in the flowering stalk, when bat pollinators have not shown up to conduct their services—as an indicator of failed pollination.[32]

Such programs and strategies to monitor and, if needed, to recover the ecosystem services provided by bats should be started very soon in the transboundary area. The stage is set for strong transboundary cooperation on behalf of bats and the ecosystem services they provide. Conservation of biodiversity for the sake of biodiversity itself has been clearly justified and has been advocated by many researchers. However, it is now clear that biodiversity conservation should go beyond simply maintaining evolutionary processes. Even if conservation of bat ecosystem services is justified by selfish reasons—human interests—it is not only right, but should be a top priority if we wish to maintain our lifestyle. It includes protecting the benefits we currently enjoy from agriculture and healthy ecosystems on both sides of the Mexico–U.S. border.

The growing demographic and political pressures, conflicting agendas, and social, political, and economic disparities of the neighboring countries on the border today place considerable additional pressure on these crucial ecosystem services.

Both the U.S. and Mexican governments historically have valued short-term (and almost in every case unsustainable) economic growth over the long-term, broad-scale, sustainable development that would benefit both countries. Short-term economic development programs are based on certain precepts of neoclassical macroeconomics, dismissing sustainable-development principles.[33] The Millennium Ecosystem Assessment has already highlighted this serious error committed around the world and indicated ways to solve this serious problem.[34]

Although transboundary conservation is incipient, in my experience federal employees on both sides of the border appear convinced of the need to seek, promote, implement, and strengthen ongoing or new transboundary collaborations. The three North American federal governments and the Trilateral Committee for Wildlife have identified bats as a top conservation priority—through the development of the North American Bat Conservation Alliance.[35] Now is the right time for such efforts; the necessary scientific justification and the political will are in place. There is still much to learn, but the critical mass, social involvement, awareness, and economic and political opportunities for successful collaboration are also in place. It is time to move forward to protect transboundary bats and their services to human well-being.

Acknowledgments

The author thanks Osiris Gaona for able technical support as well as the Consejo Nacional de Ciencia y Tecnología and the Wildlife Trust for financial support.

Notes

1. Millennium Ecosystem Assessment, *Millennium Ecosystem Assessment Reports* (Washington, D.C.: Island Press, 2005), available at http://www.millenniumassessment.org/en/Index.aspx, accessed September 24, 2009.

2. C. R. Pyke, "The implications of global priorities for biodiversity and ecosystem services associated with protected areas," *Ecology and Society* 12(1) (2007), available at http://www.ecologyandsociety.org/vol12/iss1/art4/; D. Lamb, P. D. Erskine, and J. A. Parrotta, "Restoration of degraded tropical forest landscapes," *Science* 310 (2005): 1628–32.

3. Regarding insects, see J. E. Losey and M. Vaughan, "The economic value of ecological services provided by insects," *Bioscience* 56 (2006): 311–23. For coastal ecosystems, see M. L. Martinez, A. Intralawan, G. Vazquez, O. Perez-Maqueo, P. Sutton, and R. Landgrave, "The coasts of our world: Ecological, economic, and social importance," *Ecological Economics* 63 (2007): 254–72. For soil animals, see T. Decaens, J. J. Jimenez, C. Gioia, G. J. Measey, and P. Lavelle, "The values of soil animals for conservation biology," *European Journal of Soil Biology* 42 (supplement) (2006): S23–S38. On mangroves, see M. M. P. Tognella-De-Rosa, S. R. Cunha, D. O. Lugl, M. L. G. Soares, and Y. Schaeffer-Novelli, "Mangrove evaluation—an essay," *Journal of Coastal Research* 2 (2006): 1219–24. On medicinal trees, see J. R. Figueroa and C. C. Gutierrez, "Economical valuation of trees with medicinal use in the high basin of Botanamo River, Bolivar State, Venezuela," *Interciencia* 33 (2008): 194–99. On wildlife in general, see B. P. Allen and J. B. Loomis, "Deriving values for the ecological support function of wildlife: An indirect valuation approach," *Ecological Economics* 56 (2006): 49–57.

4. N. B. Simmons, "Order Chiroptera," in *Mammal Species of the World: A Taxonomic and Geographic Reference*, 3rd ed., vol. 1, edited by D. E. Wilson and D. M. Reeder, 312–529 (Baltimore: Johns Hopkins University Press, 2005).

5. R. A. Medellín, H. T. Arita, and O. Sánchez, *Identificación de los murciélagos de México, clave de campo*, 2d ed. (Mexico City: Instituto de Ecología, Universidad Nacional Autónoma de México, 2008).

6. R. A. Medellín, "Diversity and conservation of bats in México: Research priorities, strategies, and actions," *Wildlife Society Bulletin* 31 (2003): 87–97.

7. A. M. Hutson, S. P. Mickleburgh, and P. A. Racey, *Microchiropteran Bats: Global Status Survey and Conservation Action Plan* (Gland, Switzerland: International Union for Conservation of Nature, 2001).

8. R. A. Medellín and O. Gaona, "Seed dispersal by bats and birds in forest and disturbed habitats in Chiapas, Mexico," *Biotropica* 31 (1999): 432–41.

9. U.S. Fish and Wildlife Service (FWS), *Lesser Long-nosed Bat Recovery Plan* (Albuquerque: U.S. FWS, 1995).

10. A. Casas, J. Caballero, A. Valiente-Banuet, J. A. Soriano, and P. Dávila, "Morphological variation and the process of domestication of *Stenocereus stellatus* (Cactaceae) in central Mexico," *American Journal of Botany* 86 (1999): 522–33.

11. D. H. Kelm, K. R. Wiesner, and O. von Helversen, "Effects of artificial roosts for frugivorous bats on seed dispersal in a Neotropical forest pasture mosaic," *Conservation Biology* 22 (2008): 733–41.

12. C. H. Sekercioglu, "Increasing awareness of avian ecological function," *Trends in Ecology and Evolution* 21 (2006): 464–71.

13. G. P. Nabhan and S. E. Buchman, "Services provided by pollinators," in *Nature's Services: Societal Dependence on Natural Ecosystems*, edited by G. C. Daily, 133–50 (Washington, D.C.: Island Press, 1997).

14. A. G. Valenzuela-Zapata and G. P. Nabhan, *¡Tequila! A Natural and Cultural History* (Tucson: University of Arizona Press, 2004).

15. Medellín, "Diversity and conservation of bats in México."

16. Valenzuela-Zapata and Nabhan, *¡Tequila!*

17. Medellín, "Diversity and conservation of bats in México."

18. Ibid.; see also R. A. Medellín, J. Guillermo Téllez, and J. Arroyo, "Conservation through research and education: An example of collaborative integral actions for migratory bats," in *Conservation of Migratory Pollinators and Their Nectar Corridors in North America*, edited by G. Nabhan, R. C. Brusca, and L. Holter, 43–58, Arizona-Sonora Desert Museum, Natural History of the Sonoran Desert Region no. 2 (Tucson: University of Arizona Press, 2004).

19. R. A. Medellín, "True international collaboration: Now or never," *Conservation Biology* 12 (1998): 939–40.

20. C. H. Sekercioglu, G. C. Daily, and P. R. Ehrlich, "Ecosystem consequences of bird declines," *Proceedings of the National Academy of Sciences* 101 (2004): 18042–47.

21. Medellín, "Diversity and conservation of bats in México."

22. Y. F. Lee and G. F. McCracken, "Dietary variation of Brazilian free-tailed bats links to migratory populations of pest insects," *Journal of Mammalogy* 86 (2005): 67–76.

23. T. H. Kunz, J. O. Whitaker Jr., and M. D. Wadanoli, "Dietary energetics of the insectivorous Mexican free-tailed bat *(Tadarida brasiliensis)* during pregnancy and lactation," *Oecologia* 101 (1995): 407–15.

24. C. J. Cleveland, M. Betke, P. Federico, J. D. Frank, T. G. Hallam, J. Horn, J. D. Lopez, G. F. McCracken, R. A. Medellín, A. Moreno-Valdez, C. G. Sansone, J. K. Westbrook, and T. H. Kunz, "Economic value of the pest control service provided by Brazilian free-tailed bats in south-central Texas," *Frontiers in Ecology and the Environment* 5 (2006): 238–43.

25. M. B. Kalka, A. R. Smith, and E. K. V. Kalko, "Bats limit arthropods and herbivory in a tropical forest," *Science* 320 (2008): 71.

26. P. Federico, T. G. Hallam, G. F. McCracken, S. Purucker, W. Grant, A. N. Sandoval, J. Westbrook, R. A. Medellín, C. Cleveland, C. G. Sansone, J. D. López Jr., M. Betke, A. Moreno-Valdez, and T. H. Kunz, "Brazilian free-tailed bats *(Tadarida brasiliensis)* as insect pest regulators in transgenic and conventional cotton crops," *Ecological Applications* 18 (2008): 826–37.

27. Ibid.

28. Medellín, "Diversity and conservation of bats in México"; Medellín, Guillermo Téllez, and Arroyo, "Conservation through research and education"; Cleveland et al., "Economic value of the pest control service provided by Brazilian free-tailed bats"; J. W. Horn and T. H. Kunz, "Analyzing NEXRAD doppler radar images to assess nightly dispersal patterns and population trends in Brazilian free-tailed bats *(Tadarida brasiliensis),*" *Integrative and Comparative Biology* 48 (2008): 24–39.

29. R. K. Turner and G. C. Daily, "The ecosystem services framework and natural capital conservation," *Environmental and Resource Economics* 39 (2008): 25–35.

30. J. Arroyo-Cabrales, E. K. V. Kalko, R. K. LaVal, J. E. Maldonado, R. A. Medellín, O. J. Polaco, and B. Rodríguez-Herrera, "Rediscovery of the Mexican flat-headed bat *Myotis planiceps* (Vespertilionidae)," *Acta Chiropterologica* 7 (2005): 3009–14.

31. On this network, see http://www.bioconciencia.org.mx/relcom.html.

32. R. Medellín, unpublished reports. This work is being conducted in five sites in northern Mexico.

33. D. J. Melnick, J. A. McNeely, Y. Kakabadse, G. Schmidt-Traub, and R. R. Sears, *Environment and Human Well-Being: A Practical Strategy. Achieving the Millennium Development Goals,* United Nations Millennium Project, Task Force on Environmental Sustainability Final Report (London: United Nations Development Program, EARTHSCAN, 2005).

34. Millennium Ecosystem Assessment, *Millennium Ecosystem Assessment Reports.*

35. On the Trilateral Committee, see http://www.trilat.org/annual_meetings/xiii_mtg/xiii_mtg_index_eng.html.

.nds in the Borderlands

Understanding Coupled Natural-Human Systems and Transboundary Conservation

Gerardo Ceballos, Rurik List, Ana D. Davidson, Ed L. Fredrickson, Rodrigo Sierra Corona, Lourdes Martínez, Jeff E. Herrick, and Jesús Pacheco

In a Nutshell ————————————————————————

- Grasslands in the transboundary Chihuahuan Desert provide ecosystem services that promote human well-being.
- Overgrazing, water withdrawal, drought, and prairie dog removal threaten the integrity of this important ecosystem.
- The prairie dog, a keystone species, forestalls the invasion of woody species and helps prevent desertification. However, prairie dogs are threatened, having declined by more than 95 percent throughout their historic range.
- Grassland restoration, control of shrubland expansion, and prairie dog protection are needed to maintain ecosystem services of the transboundary Chihuahuan Desert.
- Research and experience from the Janos research site in the Mexican state of Chihuahua has important implications for arid and semiarid grasslands in both the United States and Mexico.
- A new paradigm is needed to better connect agricultural production and water availability with the long-term conservation of grassland ecosystems that support these services.

Introduction

Grasslands are one of the shared ecosystems that dominate the heart of the North American continent. The Chihuahuan Desert grasslands that straddle the border between Chihuahua in Mexico and New Mexico, Arizona, and Texas in the United States are undergoing rapid transition to desertified shrubland conditions. Policy-driven land-use changes are reducing grassland capacity to maintain biodiversity and ecosystem services that support human well-being, including ranching, hydrological systems, and the prevention of desertification. Grassland degradation began first in the United States and then spread into Mexico, becoming more pronounced in parts of Mexico during the past two decades. Degradation patterns in the United States can inform Mexican conservation efforts, and relatively intact Mexican ecosystems can provide valuable insights into functional grassland systems. Research and experience from the Janos research site in the state of Chihuahua thus have important implications for arid and semiarid grassland systems that extend throughout both the United States and Mexico.

As this chapter discusses, black-tailed prairie dogs *(Cynomys ludovicianus)* in this region are a keystone species and an important ecosystem engineer critical for maintaining biodiversity and grassland ecosystem function and in turn for supporting ecosystem services to humans. This process is disrupted by poor land management (i.e., cattle overgrazing), resulting in environmental degradation that is evident in a loss of biodiversity and alteration of ecosystem function. In this chapter, we argue that conservation of grassland–prairie dog systems must be coupled with an understanding of ecosystem services and human needs. We suggest ways of overcoming the negative cycle of overgrazing and discuss how this approach can be implemented binationally.

Background

Mexico, the United States, and Canada share a continent governed by myriad linked ecological processes that reach across political boundaries. Political decisions within each country may impact ecological processes from local to global scales and transcend political boundaries, affecting a vast array of known and yet unknown ecological services. Grasslands are an important example of these linkages; they maintain a large and diverse set of living organisms and ecological processes whose persistence depends on the presence of large tracts of native grasslands that may span multiple jurisdictions. However, large-scale land-use changes caused by industrial agriculture, urbanization, infrastructure development, and desertification are reducing North American grassland systems' capacity to maintain biodiversity and ecosystem services.[1] Next, we discuss components of ecosystem change affecting the grasslands of the transboundary Chihuahuan Desert and one of its keystone species, the prairie dog.

Grassland Connectivity

The central grasslands of North America are characterized by a mosaic of grassland patches colonized and uncolonized by prairie dogs. This heterogeneous landscape supports high levels of interdependent biodiversity, often with strong associations with prairie dogs. Prairie dogs create ecological niches necessary for the persistence of many continental migratory species such as the Ferruginous Hawk *(Buteo regalis)* and the Mountain Plover *(Charadrius montanus).* The Chihuahuan grasslands, maintained by black-tailed prairie dogs (fig. 12.1), provide critical habitat for migrating birds such as Golden Eagles *(Aquila chrysaetos)* and Burrowing Owls *(Athene cunicularia).* These migratory species link this region ecologically to the northern Great Plains of the United States and Canada. More locally, populations of pronghorn antelope *(Antilocapra americana),* bison *(Bison bison)* (fig. 12.2), and other endangered species require a continuous flux of individuals between the United States and Mexico to maintain viable populations that guarantee their long-term survival. Such movement is also required as species adapt to highly variable environments typical of arid and semiarid regions.[2]

The connectivity of these dynamic ecosystems is challenged by a political boundary representing widely divergent land policies and large-scale human threats, such as global warming. Maintaining connectivity across the Mexico–U.S. border requires coordinated conservation efforts and ecologically sound management practices in both countries. The need for collaborative management is greatly increased by the wall that is being built to restrict the flow of undocumented workers across the border. For example, the border wall north of Janos, Chihuahua, is too wide and tall for large mammals, but not for people, to cross. However, because the wall is a deterrent, it funnels people to the more remote areas heavily used by wildlife. If solid walls, like those in parts of Arizona, are built in more remote areas of the Chihuahuan Desert, where currently only barbed-wire fence marks the international line, the movement of nonflying animals may cease altogether (see also chapter 6 in this volume).

Cattle Ranching

For nearly two centuries, the main economic activity in northern Mexico and the southwestern United States has been cattle ranching. During this period, poor livestock management, removal of keystone species, and suppression or lack of fire have contributed to significant landscape changes in the region.[3]

Figure 12.1. Black-tailed prairie dog. The black-tailed prairie dogs from the Janos grasslands of Chihuahua are essential for the conservation of the biodiversity of the Great Plains region of North America. Photograph by Gerardo Ceballos.

Figure 12.2. Janos bison. The Janos-Hidalgo bison of Chihuahua and New Mexico are the only free-ranging bison in Mexico and the southwestern United States. In order to persist, they need an open landscape across the international boundary. Photograph by Rurik List.

Overgrazing was widespread in the United States in the late nineteenth and early twentieth centuries;[4] however, by 1955, the situation had reversed.[5] Grazing in the United States was then being better managed, but overgrazing in Mexico was becoming more prevalent in response to federal agrarian policies that established communally managed lands called *ejidos*.

Ejido land units may be too small for effective livestock production in landscapes that are reliant on integrated ecological processes at much larger scales. For instance, in many ejidos it is often very difficult to adjust cattle numbers in response to environmental conditions; thus, the land's carrying capacity is exceeded. The increasing availability of supplemental feeds allow *ejidatarios* to maintain large numbers of cattle during drought years, and high cattle densities prevent reestablishment of palatable herbaceous species during postdrought periods. As a result, less-palatable species, most notably shrubs, have begun to replace grasslands.

In conjunction with such ranching practices, the establishment of shrub-dominated areas, the aridity of the system, and the prevalence of bare ground have caused soil erosion and degradation. The combination has had several results: loss of productive herbaceous species required for pastoralist cultures; loss of key grassland species; and changes in ecological processes, such as fire, that are necessary for maintaining the structure and function of the grassland ecosystem. Many people, especially young people, have left the region's ejidos for the United States because the now degraded landscapes no longer support the ejidatarios' agricultural livelihoods.

The Janos region of northern Mexico, at the northwestern corner of the state of Chihuahua, is a top priority for biodiversity conservation in North America. However, pressures from agricultural development as well as inadequate management practices and planning seriously threaten the region's biodiversity. Conservation efforts have often failed by neglecting to incorporate people as part of the ecological system. The success of conservation programs and economic development will be limited if efforts are focused on restricting natural-resource use rather than on improving current management practices. Economic activities such as agriculture and grazing can benefit from greater understanding of the region's ecological setting, while becoming compatible with conservation.

The Grassland–Prairie Dog Ecosystem

The conflict between the conservation of grassland ecosystems that support prairie dogs and economic activities in the southwestern United States and northern Mexico is a case study in the problems mentioned in the previous section. It is clear from our research in the Janos region that declines in prairie dog populations have their own long-term, large-scale, negative effects on biodiversity, ecosystem services, economic activities, and human well-being. Coupling the conservation of prairie dogs with economic activities is perhaps the only way to maintain the livelihood of local people and regional economies while preserving the ecological characteristics that made these activities possible. In the following paragraphs, we describe in detail the complex ecological setting and its direct relationships with human well-being. This new way of looking at these relationships makes economic development, conservation, and human well-being more compatible.

Prairie dogs are keystone species and ecosystem engineers of the Chihuahuan Desert grasslands.[6] They transform grassland landscapes by grazing and constructing extensive burrow systems. Their mounds and large colonies create a mosaic of islandlike habitat patches that differ in biotic composition and ecosystem properties from surrounding grassland areas, increasing biodiversity across the greater grassland landscape.[7] Many other grassland species utilize their burrows and colonies as key habitat and depend on prairie dogs as prey.[8] Prairie dogs play an important role in cycling nutrients and preventing shrub encroachment.[9] Such ecosystem services are especially valuable in the Chihuahuan Desert grasslands, where shrub encroachment and desertification are major issues.[10]

Despite their ecological importance, prairie dog populations are declining throughout their range;[11] all five species (Gunnison *Cynomys gunnisoni*, white-tailed *C. leucurus*, black-tailed *C. ludovicianus*, Mexican *C. mexicanus*, and Utah *C. parvidens*) are listed or are the subject of petitions for listing as threatened or endangered in either the United States or Mexico. Threats to prairie dogs in the United States include sylvatic plague, urban development, agriculture, poisoning, and shooting. The prevailing threats in Mexico are desertification, agriculture, and, possibly, climate change (predicted to cause an increased frequency and severity of drought in the Chihuahuan Desert region).[12] Prairie dog declines in the United States have caused declines in associated species, including the black-footed ferret *(Mustela nigripes)*, Burrowing Owl, Ferruginous Hawk, Mountain Plover, and swift fox *(Vulpes velox)*.[13] Over a thirteen-year period, our research in the Janos region has demonstrated drastic reductions in biodiversity along with declining prairie dog populations (fig. 12.3). These losses can be attributed to extreme land degradation due to overgrazing and conversion of native grassland to cropland, combined with natural drought cycles. The threefold decline in prairie dog densities and the approximately 70 percent reduction

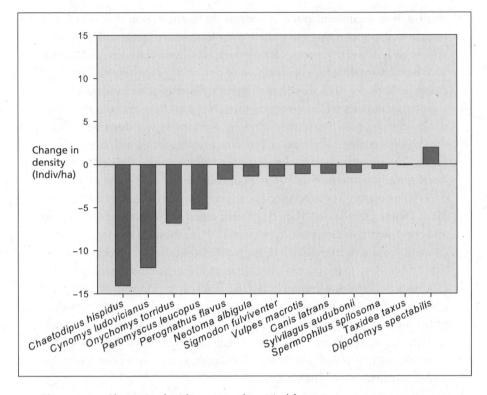

Figure 12.3. Change in abundance over the period from 1992–1994 to 2000–2001 of mammal species known to be associated with prairie dogs and their colonies in the Janos region of Chihuahua.

in prairie dog colony size[14] have undoubtedly contributed to the decline of many species known to utilize prairie dogs as prey or their colonies for habitat, such as the coyote *(Canis latrans)*, kit fox *(Vulpes macrotis)*, badger *(Taxidea taxus)*, and other small mammals, and are further jeopardizing the viability of the black-footed ferret population in the region.

Today, the most significant threat to the persistence of prairie dogs in Mexico (including the Mexican prairie dog, *Cynomys mexicanus*, in the eastern Chihuahuan Desert) is the expansion of irrigated agriculture for cotton, potato, and corn production. Despite a ban on developing new groundwater wells in Janos and the requirement for environmental impact assessments to convert native vegetation to agriculture, some farmers illegally drill wells and plow over grasslands and prairie dog towns. Because of corruption of officials and the inefficient enforcement of environmental laws, regulatory mechanisms have proved insufficient to stop illegal well drilling and land conversion.

Cattle and Prairie Dogs

Prairie dogs have been exterminated throughout an extensive portion of their historic range under the assumption that they compete with cattle for forage. Depending on management and environmental conditions, the interaction between cattle and prairie dogs can be either synergistic or competitive. Synergistic interactions between ungulates (a category including cattle) and prairie dogs occur in areas of the mixed grass prairie of North America. It has been documented that large ungulates, such as bison and elk, form "grazing associations" with prairie dogs. Both native and domestic ungulates are often attracted to prairie dog colonies because the forage there is more nutritious. In turn, prairie dogs benefit when livestock help to maintain lower-stature vegetation that allows greater visibility of and vigilance against predators.[15]

Without the presence of prairie dogs, cattle can cause grasslands to be transformed into shrublands. Ungulates eat the seed pods of the mesquite, a shrub that is replacing grasslands. They then deposit the seeds on the soil, often far from the parent plant. If the seeds germinate, and the seedlings survive and mature, they produce more seed. Plant by plant, mesquite eventually replaces grasses and the amount of forage for ungulates steadily declines, the structure of the vegetation is altered, and the whole system drifts to desertification. However, if prairie dogs are present, they limit the establishment of woody vegetation and maintain grassland systems by removing pods, eating mesquite seedlings and saplings, and clipping shrubs.[16] Prairie dogs, therefore, control mesquite expansion and preserve the grasses for both prairie dogs and ungulates to eat.

The presence or absence of large prairie dog colonies in northern Mexico is also related to desertification processes as well as to the ranching practices and grassland transformation discussed here. With the ejido system and the 1970s agrarian movement in Mexico, overgrazing by cattle led to competition between prairie dogs and cattle for the same grass; this competition worsened with drought in the 1990s and early 2000s. In the 1980s, owners of private land began poisoning prairie dogs, allowing mesquite shrubs to gain dominance. In communal lands, the poisoning was less intensive. Instead of the grassland's being transformed to shrubland in the absence of prairie dogs, the area was transformed from a perennial grassland to an annual grassland and bare ground.

Recent studies in Chihuahua demonstrate that the persistence of prairie dogs has been a key element in maintaining grasslands even under severe overgrazing and drought. In areas where prairie dogs have disappeared, mesquite has rapidly invaded. When prairie dog colonies have been restored, they exterminate mesquite.

In the southwestern United States, as in Mexico, desertification of grasslands is in part related to prairie dog disappearance.[17] The cattle boom of the 1880s and the U.S. government–funded poisoning campaigns eradicated all but a few U.S. prairie dog colonies. Mesquite expanded in areas once dominated by prairie dogs, and grasslands began to disappear, leading to desertification, such as that observed at the La Jornada Experimental Range in southern New Mexico.

Ecosystem Services and Change

Grassland loss and conservation have significant implications for many ecosystem services. During the past 150 years, tree- and shrub-dominated plant communities have replaced grasslands throughout much of the Mexico–U.S. borderlands region.[18] Patterns and rates of shrub invasion vary with soil, topography, climate, weather, and disturbance history, in addition to changes in the populations and population dynamics of native animals. Although the relative importance of these factors is unknown, several general trends and patterns are widely recognized throughout the region. Creosote bush *(Larrea tridentata)*, tarbush *(Fluorensia cernua)*, and mesquite *(Prosopis* spp.) have expanded into grasslands at lower elevations in the Chihuahuan Desert.[19] Mesquite dominates most former black grama grasslands *(Bouteloua eriopoda)* on sandy soils but is also increasingly found on finer-textured soils. Similar transitions have occurred in the Sonoran Desert with a different mix of species and a greater dominance by succulent species. As demonstrated earlier, the transformations are accelerated by and may be initiated by overgrazing and drought,[20] although climate change, changes in human use, and the eradication of keystone species such as prairie dogs also play a role.[21]

The shift from grasslands to mesquite-dominated shrublands has dramatic effects on ecosystem services. This transformation of grassland to shrubland changes ecosystem processes and the capacity to support ecosystem services. Although the specific nature of the changes varies with plant community, soils, and landscape position, the general relationships are similar throughout the region. Grassland processes in arid and semiarid lands are related to four types of ecosystem services identified by the Millennium Ecosystem Assessment: water quality, air quality, food and fiber, and biodiversity.[22] In particular, with the shift from grasslands to mesquite-dominated shrublands,

1. *Water quality* is enhanced in seasonal streams because there is less erosion from shrublands.[23] This process can also increase long-term water quantity available for human use because reservoir siltation is slowed. *Water availability* for humans and wildlife is modified because grasslands increase infiltration and reduce runoff relative to shrublands. This process can increase the duration of perennial stream flow and spring production, although a recent review indicates that these benefits are more likely to occur in relatively higher-precipitation zones.[24]

2. *Air quality* is maintained because grasslands have smaller plant interspaces, which limit wind erosion. This erosion is due primarily to the change in spatial distribution, with larger plant patches and interspaces. In addition to its effects on human health, erosion can reduce the aesthetic value of many landscapes.

3. The potential production of *food and fiber* is generally greater in grasslands through both domesticated livestock and wildlife. Long-term data from the Jornada Experimental Range in New Mexico indicate that grasslands at least have the potential to support greater rates of net primary production than shrublands[25] and that herbaceous forage production is uniformly higher in grasslands.

4. *Biodiversity* is generally thought to be better supported by grasslands, in part because plant diversity is often greater[26] and because prairie dog towns maintain more species, including species at risk, than unoccupied grasslands, and many of those species are not found in shrublands. As a consequence, grassland loss is increasingly associated with regional declines in biodiversity.

Although most ecosystem services currently valued by people in the border region are more effectively supplied by grasslands, mesquite-dominated communities do provide some critical services. Bird diversity is similar or even higher in shrub-dominated systems due to greater habitat structure.[27] Honey from mesquite shrubs located in areas with shallow water tables is valued for its unique flavor and often sells at a premium of 50 percent or more higher than honey derived from other sources. Mesquite beans also have value for both humans and wildlife.[28]

Nonetheless, mesquite dominance reduces future management options and therefore the ability of managers—including ranchers, governments, and conservation organizations—to manage for different ecosystem services. It is relatively easy to replace grasses with shrubs, but it is difficult to establish or reestablish perennial grasses once a grassland has been transformed into a shrubland. For example, maximizing forage-production services for livestock in shrublands cannot be accomplished without significant external inputs.[29] These "thresholds" beyond which it is difficult to return a system to its original state are typical of arid and semiarid ecosystems and are associated with the fundamental changes in the soil-plant-animal relationships and feedbacks discussed in this chapter.[30] To predict and prevent such thresholds, both conservation biologists and land managers should use models and decision-making tools that integrate ecosystem processes, such as soils, with wildlife feedbacks.[31] These models can be used to communicate current knowledge about grass–shrub transitions and identify knowledge gaps to improve land management.

Approach to Grasslands Management

Over the past twenty years, the Janos region has become a premier laboratory for understanding the importance of coupled human-ecological systems in achieving grassland-conservation goals. Our research in this region in northwestern Chihuahua, along with related work by the Malpai Borderlands Group in southern New Mexico and Arizona, strongly demonstrates the need to understand and maintain a complete suite of ecosystem processes in order to maintain viable human populations, functional ecological systems, and ecosystem services.

The Mexican government is in the process of designating the Janos region for protection as a half-million-hectare Biosphere Reserve; this pending designation provides the foundation for conservation of the region's rich biodiversity. However, to be effective, conservation requires proper management of rural economic activities such as grazing, hunting, and agriculture. Adequate management requires techniques for productive activities that take into account the region's ecology and sound ecological zoning. Several key management issues are involved in coupling economic and human development with conservation of biodiversity.

Grassland Restoration

Restoring perennial grasslands is critical for the conservation of Chihuahuan Desert ecosystems. Grassland restoration in the southwestern United States and northern Mexico can benefit from the establishment of large prairie dog colonies and better management of cattle grazing, along with a consideration of the effects of climate change. The prairie dog poisoning promoted by the U.S. government needs to end, and the funds used for it should be reallocated to subsidize ranchers willing to have prairie dogs on their lands. Lessons learned by the Malpai Borderlands Group in southern New Mexico and Arizona and from our work in the Janos site should be applied to other regions to start a large-scale restoration of grassland ecosystems.

For example, fire and prairie dogs play important roles in eliminating seedlings of desert shrubs such as mesquite; both large grazers and fire promote the expansion of prairie dog populations by reducing vegetation height. The elimination of natural fires, extermination of prairie dogs, and overgrazing have promoted the expanse of mesquite shrublands. The semiarid grasslands of North America evolved with large herbivores such as bison; in the latter's absence, well-managed grazing by cattle may in fact have a positive influence on grasslands while also promoting control of mesquite and contributing to local economies, but only if prairie dogs are included in the mix.

Outreach

Strong efforts should be made to conduct outreach both locally and regionally. In both Mexico and the United States, it will be critical to educate landowners and authorities about the role of prairie dogs in maintaining the grasslands, the compatibility of prairie dog and cattle grazing, and the effects of drought and overgrazing in promoting desertification. These issues are most likely the most critical for the persistence of large tracts of perennial grasslands in the Chihuahuan Desert. Such outreach efforts, however, should be conducted differently in the two countries. In the United States, outreach has to be concentrated on both the government and private landowners. In Mexico, it has to address both caretakers of communal lands and private landowners. In both countries, it has to emphasize the benefits of maintaining prairie dogs for the grasslands and the compatibility of conservation and human activities.

Conclusions

Prairie dogs are essential to maintaining grassland ecosystems, ecosystem services, and human well-being. First, the conservation status of borderland grassland–prairie dog systems has important transboundary implications for biodiversity. Prairie dogs support local, regional, and continental biodiversity by maintaining open grasslands and providing habitat for many species.

For instance, the status of migratory bird and raptor breeding populations 1,000 kilometers (620 miles) north of the Mexico–U.S. border can be linked to the health of grasslands in the narrow borderland region. For species such as bison and pronghorn, the influence is more local, and the need of open grasslands to maintain viable populations reaches only a few tens of kilometers north and south of the international boundary. In addition, by maintaining open grasslands, prairie dogs support a more productive ecosystem for cattle, help to increase the amount of water percolating to the water table, and suppress woody plants, thereby curbing desertification and helping to maintain soil stability.

Our results from Janos show that cattle ranching and conservation of prairie dogs are compatible. The lessons learned in the grasslands of the borderlands region of Mexico can be applied to many of the nearly 20 million hectares (50 million acres) that prairie dogs formerly occupied in North America, where cattle ranching is today the main economic activity.

The trilateral (United States, Mexico, Canada) Commission for Environmental Cooperation has published the Black-Tailed Prairie Dog Conservation Action Plan. For the plan to have an impact, the commission will need to coordinate the work of different actors involved in prairie dog conservation across their distribution range in North America. Funding will be needed to combat eradication programs, increase research, and develop improved land-management techniques in priority conservation areas.

Grasslands have played a major role in human history, from the origin of agriculture and the establishment of important past and present cultures to the production of food for a large part of humanity. We have a moral obligation to do whatever is in our power to ensure that the ecological processes that shaped the characteristically rich biodiversity of the grasslands of North America persist long into the future.

Acknowledgments

We thank Laura López-Hoffman for inviting us to contribute to this book and two anonymous reviewers whose help improved the chapter. Funding for our work was provided by the Universidad Nacional Autónoma de México, the J. M. Kaplan Fund, the Whitley Fund for Nature, The Nature Conservancy, and the Comisión Nacional para el Conocimiento y Uso de la Biodiversidad.

Notes

1. J. Foley, R. DeFries, G. P. Asner, C. Barford, G. Bonan, S. R. Carpenter, F. S. Chapin, M. T. Coe, G. C. Daily, H. K. Gibbs, J. H. Helkowski, T. Holloway, E. A.

Howard, C. J. Kucharik, C. Monfreda, J. A. Patz, I. C. Prentice, N. Ramankutty, and P. K. Snyder, "Global consequences of land use," *Science* 309 (2005): 570–74; L. F. Huenneke, J. P. Anderson, M. Remmenga, and W. H. Schlesinger, "Desertification alters patterns of aboveground net primary production in Chihuahuan ecosystems," *Global Change Biology* 8 (2002): 247–64.

2. G. Ceballos, J. Pacheco, R. List, P. Manzano-Fischer, G. Santos, and M. Royo, "Prairie dogs, cattle, and crops: Diversity and conservation of the grassland-shrubland mosaics in northwestern Chihuahua, Mexico," in *Biodiversity, Ecosystems, and Conservation in Northern Mexico*, edited by J-L. E. Cartron, G. Ceballos, and R. S. Felger, 1–19 (Oxford, U.K.: Oxford University Press, 2005); R. List, "The impacts of the border fence on wild mammals," in *A Barrier to Our Shared Environment: The Border Wall Between Mexico and the United States*, edited by A. Córdova and C. A. de la Parra, 77–86 (Mexico City and San Diego: Secretaría de Medio Ambiente y Recursos Naturales, Instituto Nacional de Ecología, El Colegio de la Frontera Norte, and Southwest Consortium for Environmental Research and Policy, 2007).

3. T. L. Fleischner, "Ecological costs of livestock grazing in western North America," *Conservation Biology* 8 (1994): 629–44.

4. A. Leopold, "Conservationist in Mexico," *American Forests* 43 (1937): 118–21.

5. R. B. Villa, "Observaciones acerca de la última manada de berrendos *(Antilocapra americana)* en el estado de Chihuahua, México," *Anales del Instituto de Biología* 26 (1955): 229–36.

6. G. Ceballos, J. Pacheco, and R. List, "Influence of prairie dogs *(Cynomys ludovicianus)* on habitat heterogeneity and mammalian diversity in Mexico," *Journal of Arid Environments* 41 (1999): 161–72; A. D. Davidson and D. C. Lightfoot, "Keystone rodent interactions: Prairie dogs and kangaroo rats structure the biotic composition of a desertified grassland," *Ecography* 29 (2006): 755–56; N. B. Kotliar, B. J. Miller, R. P. Reading, and T. W. Clark, "The prairie dog as a keystone species," in *Conservation of the Black-tailed Prairie Dog: Saving North America's Western Grasslands*, edited by J. L. Hoogland, 53–64 (Washington, D.C.: Island Press, 2006); A. D. Davidson and D. C. Lightfoot, "Burrowing rodents increase landscape heterogeneity in a desert grassland," *Journal of Arid Environments* 72 (2008): 1133–45.

7. Davidson and Lightfoot, "Keystone rodent interactions"; A. D. Whicker and J. K. Detling, "Ecological consequences of prairie dog disturbances: Prairie dogs alter grassland patch structure, nutrient cycling, and feeding-site selection by other herbivores," *Bioscience* 38 (1988): 778–85.

8. Ceballos, Pacheco, and List, "Influence of prairie dogs"; Kotliar et al., "The prairie dog as a keystone species"; R. List, P. Manzano-Fischer, and D. W. Macdonald, "Coyote and kit fox diets in a prairie dog complex in Mexico," in *The Swift Fox: Ecology and Conservation of Swift Foxes in a Changing World*, edited by M. Sovada and L. Carbyn, 183–88 (Regina: Canadian Plains Research Center, University of Regina, 2003); M. V. Lomolino and G. A. Smith, "Terrestrial vertebrate communities at black-tailed prairie dog *(Cynomys ludovicianus)* towns," *Biological*

Conservation 115 (2003): 89–100; A. D. Davidson and D. C. Lightfoot, "Interactive effects of keystone rodents on the structure of desert grassland arthropod communities," *Ecography* 30 (2007): 515–25.

9. Whicker and Detling, "Ecological consequences of prairie dog disturbances"; J. F. Weltzin, S. L. Dowhower, and R. K. Heitschmidt, "Prairie dog effects on plant community structure in southern mixed-grass prairie," *The Southwestern Naturalist* 42 (1997): 251–58.

10. W. H. Schlesinger, J. F. Reynolds, G. L. Cunningham, L. F. Huenneke, W. M. Jarrell, R. A. Virginia, and W. G. Whitford, "Biological feedbacks in global desertification," *Science* 247 (1990): 1043–48.

11. B. Miller, G. Ceballos, and R. Reading, "The prairie dog and biotic diversity," *Conservation Biology* 8 (1994): 677–81; Hoogland, *Conservation of the Black-tailed Prairie Dog.*

12. Ceballos et al., "Prairie dogs, cattle, and crops"; Hoogland, *Conservation of the Black-tailed Prairie Dog.*

13. Kotliar et al., "The prairie dog as a keystone species."

14. G. Ceballos, E. Mellink, and L. R. Hanebury, "Distribution and conservation status of prairie dogs *Cynomys mexicanus* and *Cynomys ludovicianus* in Mexico," *Biological Conservation* 63 (1993): 105–12.

15. D. L. Coppock, J. E. Ellis, J. K. Detling, and M. I. Dyer, "Plant-herbivore interactions in a North American mixed-grass prairie. II. Responses of bison to modification of vegetation by prairie dogs," *Oecologia* 56 (1983): 10–15; K. Krueger, "Feeding relationships among bison, pronghorn, and prairie dogs: An experimental analysis," *Ecology* 67 (1986): 760–70; B. J. Miller, R. P. Reading, D. E. Biggins, J. E. Detling, S. E. Forrest, J. L. Hoogland, J. Javersak, R. D. Miller, J. Proctor, J. Truett, and D. W. Uresk, "Prairie dogs: An ecological review and current biopolitics," *Journal of Wildlife Management* 71 (2007): 2801–10.

16. J. F. Weltzin, S. Archer, and R. K. Heitschmidt, "Small-mammal regulation of vegetation structure in a temperate savanna," *Ecology* 78 (1997): 751–63; Weltzin, Dowhower, and Heitschmidt, "Prairie dog effects on plant community structure in southern mixed-grass prairie."

17. Weltzin, Archer, and Heitschmidt, "Small-mammal regulation of vegetation structure in a temperate savanna"; Weltzin, Dowhower, and Heitschmidt, "Prairie dog effects on plant community structure in southern mixed-grass prairie."

18. L. C. Buffington and C. H. Herbel, "Vegetational changes on a semidesert grassland range from 1858 to 1963," *Ecological Monographs* 35 (1965): 139–64; D. D. Breshears, N. S. Cobb, P. M. Rich, K. P. Price, C. D. Allen, R. G. Balice, W. H. Romme, J. H. Kastens, M. L. Floyd, J. Belnap, J. J. Anderson, O. B. Myers, and C. W. Meyer, "Regional vegetation die-off in response to global-change-type drought," *Proceedings of the National Academy of Sciences* 102 (2005): 15144–48; D. P. C. Peters and R. P. Gibbens, "Plant communities in the Jornada basin: The dynamic landscape," in *Structure and Function of a Chihuahuan Desert Ecosystem*, edited by K. M. Havstad, L. F. Huenneke, and W. H. Schlesinger, 211–31, Jornada Basin Long-Term Ecological Research Site (Oxford, U.K.: Oxford University Press, 2006).

19. Buffington and Herbel, "Vegetational changes on a semidesert grassland range from 1858 to 1963"; Peters and Gibbens, "Plant communities in the Jornada basin."

20. Schlesinger et al., "Biological feedbacks in global desertification."

21. E. L. Fredrickson, R. E. Estell, A. Laliberte, and D. M. Anderson, "Mesquite recruitment in the Chihuahuan Desert: Historic and prehistoric patterns with long-term impacts," *Journal of Arid Environments* 65 (2006): 285–95.

22. Millennium Ecosystem Assessment, *Millennium Ecosystem Assessment Reports* (Washington, D.C.: Island Press, 2005), available at http://www.millenniumassessment.org/en/Index.aspx; D. P. C. Peters, R. A. Pielke Sr., B. T. Bestelmeyer, C. D. Allen, S. Munson-McGee, and K. M. Havstad, "Cross-scale interactions, non-linearities, and forecasting catastrophic events," *Proceedings of the National Academy of Sciences* 41 (2004): 15130–35; K. M. Havstad, D. C. Peters, R. Skaggs, J. Brown, B. T. Bestelmeyer, E. L. Fredrickson, J. E. Herrick, and J. Wright, "Ecosystem services to and from rangelands of the western United States," *Ecological Economics* 64 (2007): 261–68.

23. A. D. Abrahams, M. Neave, W. H. Schlesinger, J. Wainright, D. A. Howes, and A. J. Parsons, "Biogeochemical fluxes across piedmont slopes of the Jornada basin," in Havstad, Huenneke, and Schlesinger, eds., *Structure and Function of a Chihuahuan Desert Ecosystem*, 150–75.

24. B. P. Wilcox and T. L. Thurow, "Emerging issues in rangeland ecohydrology: Vegetation change and the water cycle," *Rangeland Ecology and Management* 59 (2006): 220–24.

25. L. F. Huenneke and W. H. Schlesinger, "Patterns of net primary production in Chihuahuan Desert ecosystems," in Havstad, Huenneke, and Schlesinger, eds., *Structure and Function of a Chihuahuan Desert Ecosystem*, 232–46.

26. Peters and Gibbens, "Plant communities in the Jornada basin."

27. W. G. Whitford, "Desertification and animal biodiversity in the desert grasslands of North America," *Journal of Arid Environments* 37 (1997): 709–20.

28. E. Fredrickson, K. M. Havstad, and R. Estell, "Perspectives on desertification: South-western United States," *Journal of Arid Environments* 39 (1998): 191–207.

29. B. A. Roundy and S. H. Biedenbender, "Revegetation in the desert grassland," in *The Desert Grassland*, edited by M. P. McClaran and T. R. Van Devender, 265–303 (Tucson: University of Arizona Press, 1995).

30. B. T. Bestelmeyer, J. R. Brown, K. M. Havstad, R. Alexander, G. Chavez, and J. Herrick, "Development and use of state-and-transition models for rangelands," *Journal of Range Management* 56(2) (2003): 114–26; T. K. Stringham, W. C. Krueger, and P. L. Shaver, "State and transition modeling: An ecological process approach," *Journal of Range Management* 56(2) (2003): 106–13.

31. J. E. Herrick, B. T. Bestelmeyer, S. Archer, A. Tugel, and J. R. Brown, "An integrated framework for science-based arid land management," *Journal of Arid Environments* 65 (2006): 319–35.

How a Changing Climate Will Affect U.S.–Mexico Transboundary Conservation

Matt Skroch

Western North America, in particular the southwestern United States and much of Mexico, is experiencing a trend toward a warmer, more arid climate.[1] This climatic shift will have serious implications for transboundary conservation both in the U.S.–Mexico borderlands and along the north–south corridors traveled by migratory species.

Climatic Prediction Summary

The Intergovernmental Panel on Climate Change (IPCC) predicts that average annual temperature in the U.S.–Mexico borderlands, in addition to much of western North America and Central America, will likely increase by about 1.5°C by 2030 to more than 3.0°C by 2100.[2] In combination with this warming trend, annual precipitation is projected to decrease across the region.[3] In winter months, the polar jet stream—responsible for delivering moist polar maritime air masses over the U.S. Southwest—is expected to track farther north, increasing the frequency of high-pressure systems that fuel hot, dry weather.[4] The IPCC predicts a 10 to 15 percent reduction in winter rains by 2050, although downscaled climate models disagree.[5] Summer precipitation, created by the northward movement of tropical maritime air masses over the Gulf of Mexico and Gulf of California, will likely remain static or slightly increase due to higher ocean temperatures.[6] Winter snowpack in the headwaters of the Colorado River and the Rio Grande will be reduced, leading to decreased spring runoff.[7]

Water Balance, Availability, and Use in the Borderlands

Due to lower mean annual precipitation and reduced snowpack in the headwaters of the Colorado and Rio Grande systems, less water will be available to fill reservoirs.[8] Some researchers predict 20 to 40 percent less runoff in the Colorado River basin by 2060,[9] and other model results suggest a 50 percent chance that both

Lake Mead and Lake Powell will drop below minimum levels needed for power production by 2017.[10] Increasing groundwater extraction to compensate for less surface water will not be a long-term option because many aquifers within the U.S.–Mexico transboundary region have already been degraded or drawn down.[11] These results will have far-ranging impacts on ecological and human-based water systems. For instance, a 2°C increase in annual mean temperature has been shown to reduce water availability for lactating fringed myotis bats *(Myotis thysanodes)* by almost 60 percent during their critical reproductive period.[12]

Large-Scale Vegetation Shifts and Invasive Species in the Borderlands

Due to increased annual temperatures, current vegetation associations may change. Research has suggested a northern and eastern expansion of the Sonoran Desert. Frost-intolerant subtropical species, no longer inhibited by freezing temperatures, may colonize more northern latitudes.[13] The exotic species buffelgrass *(Pennisetum ciliare)* will concurrently continue to transform large portions of the Sonoran Desert into an African grassland savanna.[14] As temperatures rise, the grass will likely spread upward in elevation, significantly altering the ecological structure and function of those areas. At higher elevations, drought and extreme temperatures will likely continue to cause large vegetation die-offs, and areas will develop different vegetation associations than previously present.[15] The continuation of severe drought conditions, coupled with invasive species, is expected to increase the frequency and severity of wildland fires.[16] These fires will create positive feedback loops toward increased regional aridity and large-scale vegetation shifts: more aridity and fires set up conditions for more aridity and fires.

Continental Impacts: Missed Migratory Connections?

From a larger continental and intercontinental perspective, climate change is disrupting shifts in the timing of life-cycle events (phenology) between migrating wildlife species and the plants they depend on (e.g., flowering or seed-production timing for food). This potential shift is particularly critical for the north–south continental routes that migratory species travel because regional climate shifts might not be synchronized with changes in other parts of the Americas.[17] For instance, the IPCC indicates increasingly different precipitation trends for western North America and central Mexico (decreasing precipitation) as compared to the Intertropical Convergence Zone (increasing precipitation) over northwestern South America.[18] Wildlife migration may fall out of sync with the timing of critical plant resources they need during their migration. For example, the monarch butterfly *(Danaus plexippus)* uses well-defined habitat niches with specific host plants during winter and summer migrations; if the habitat niches shift northward as predicted, the migrating butterflies might not overlap with their habitats and host plants.[19] Likewise, the seasonal arrival times of migratory songbirds are influenced by temperature and large-scale shifts in sea-surface temperature and atmospheric circulation patterns, such as the El Niño–Southern Oscillation. If the dates of songbird migrations no longer coincide with the growth or blooming of the plant food sources they depend on, songbird populations might decline.[20]

Notes

1. R. Seager, M. F. Ting, I. M. Held, Y. Kushnir, J. Lu, G. Vecchi, H-P. Huang, N. Harnik, A. Leetmaa, N-C. Lau, C. Li, J. Velez, and N. Naik, "Model projections of an imminent transition to a more arid climate in southwestern North America," *Science* 316(5828) (2007): 1181–84.

2. S. Solomon, G-K. Plattner, R. Knutti, and P. Friedlingstein, "Irreversible climate change due to carbon dioxide emissions," *Proceedings of the National Academy of Sciences* 106(6) (2009): 1704–9; J. H. Christensen, B. Hewitson, A. Busuioc, A. Chen, X. Gao, I. Held, R. Jones, R. K. Kolli, W-T. Kwon, R. Laprise, V. Magaña Rueda, L. Mearns, C. G. Menéndez, J. Räisänen, A. Rinke, A. Sarr, and P. Whetton, "Regional climate projections," in *Climate Change 2007: The Physical Science Basis. Contribution of Working Group I to the Fourth Assessment Report of the Intergovernmental Panel on Climate Change*, edited by S. Solomon, D. Qin, M. Manning, Z. Chen, M. Marquis, K. B. Averyt, M. Tignor, and H. L. Millerm, 847–940 (Cambridge, U.K.: Cambridge University Press, 2007).

3. Christensen et al., "Regional climate projections."

4. Ibid.; D. J. Seidel, Q. Fu, W. J. Randel, and T. J. Reichler, "Widening of the tropical belt in a changing climate," *Nature Geoscience* 1 (2008): 21–24.

5. Christensen et al., "Regional climate projections"; N. S. Christensen and D. P. Lettenmaier, "A multimodel ensemble approach to assessment of climate change impacts on the hydrology and water resources of the Colorado River basin," *Hydrology and Earth System Sciences* 11 (2007): 1417–34.

6. Christensen et al., "Regional climate projections."

7. I. T. Stewart, "Changes in snowpack and snowmelt runoff for key mountain regions," *Hydrological Processes* 23(1) (2009): 78–94.

8. T. P. Barnett and D. W. Pierce, "Sustainable water deliveries from the Colorado River in a changing climate," *Proceedings of the National Academies of Science* 106(18) (2009): 7334–38.

9. P. C. D. Milly, K. A. Dunne, and A. V. Vecchia, "Global pattern of trends in streamflow and water availability in a changing climate," *Nature* 438(7066) (2005): 347–50.

10. Barnett and Pierce, "Sustainable water deliveries from the Colorado River in a changing climate."

11. Comisión Nacional del Agua, *Statistics on Water in Mexico* (Mexico City: Comisión Nacional del Agua, 2007).

12. R. A. Adams and M. A. Hayes, "Water availability and successful lactation by bats as related to climate change in arid regions of western North America," *Journal of Animal Ecology* 77(6) (2008): 1115–21.

13. J. L. Weiss and J. T. Overpeck, "Is the Sonoran Desert losing its cool?" *Global Change Biology* 11(12) (2005): 2065–77.

14. K. A. Franklin, K. Lyons, P. L. Nagler, D. Lampkin, E. P. Glenn, F. Molina-Freaner, T. Markow, and A. R. Huete, "Buffelgrass *(Pennisetum ciliare)* land conversion and productivity in the plains of Sonora, Mexico," *Biological Conservation* 127(1) (2006): 62–71.

15. D. D. Breshears, N. S. Cobb, P. M. Rich, K. P. Price, C. D. Allen, R. G. Balice, W. H. Romme, J. D. Kastens, M. L. Floyd, J. Belnap, J. J. Anderson, O. B. Myers, and C. W. Meyer, "Regional vegetation die-off in response to global-change-type drought," *Proceedings of the National Academy of Sciences* 102 (2005): 15144–48.

16. D. B. Fagre, D. L. Peterson, and A. E. Hessl, "Taking the pulse of mountains: Ecosystem responses to climatic variability," *Climatic Change* 59(1–2) (2003): 263–82.

17. R. V. Batalden, K. Oberhauser, and A. T. Peterson, "Ecological niches in sequential generations of eastern North American monarch butterflies (Lepidoptera:Danaidae): The ecology of migration and likely climate change implications," *Environmental Entomology* 36(6) (2007): 1365–73; D. P. Macmynowski, T. L. Root, G. Ballard, and G. Geupel, "Changes in spring arrival of Nearctic-Neotropical migrants attributed to multiscalar climate," *Global Change Biology* 13(11) (2007): 2239–51.

18. Christensen et al., "Regional Climate Projections."

19. Batalden, Oberhauser, and Peterson, "Ecological niches in sequential generations of eastern North American monarch butterflies."

20. B. Huntley, Y. C. Collingham, S. G. Willis, and R. E. Green, "Potential impacts of climatic change upon geographical distributions of birds," *Ibis* 148 (2006): 8–28.

Border Security
and Conservation

Border Security and Conservation

U.S. federal actions to enhance border security are a relatively recent but significant driver of environmental change in the U.S.–Mexico borderlands. These policies are impacting transboundary species and spaces in at least two discernable ways. First, they have changed the movement patterns of people, smuggling, and enforcement activities through the border region. Since the mid-1990s, they have funneled these activities away from the border's urban areas and into its remote landscapes and protected areas. Second, through the construction of barriers and enforcement infrastructure along significant segments of the boundary—increasingly in rural areas— they have sliced migratory corridors for many animals in two.

The issue of border security appears throughout this volume, speaking to its effects on a variety of shared species, spaces, and ecosystem services. The topic therefore has warranted the deeper examination that the three chapters in this section provide. These chapters critique how current security policies have impacted not only border landscapes, but also the capacity for collaborative conservation in the borderlands.

Chapter 13, by Jessica Piekielek, analyzes the tension between unilateral U.S. security measures and the cooperative conservation efforts by U.S. and Mexican "sister parks"—adjacent U.S. and Mexican protected areas—and their staff. It specifies the ways in which recent security dilemmas have posed obstacles to collaborative conservation between sister parks in Arizona and Sonora, and reaffirms the need for protected-area natural-resource managers to continue their cooperative efforts.

In chapter 14, Christopher Sharp and Randy Gimblett document the direct and indirect impacts of the funneling of traffic through protected areas by looking at the evidence of increased activity throughout Organ Pipe Cactus National Monument. They show that distinct usage vectors— including migrants, smugglers, law enforcement, park staff, and recreation visitors—have interrelated and cumulative impacts on the monument's biophysical attributes.

In chapter 15, Brian Segee and Ana Córdova chronicle U.S. border security legislation since 2005. They consider how the barriers and the waivers of environmental regulations are affecting environments and disrupting wildlife migration corridors along the entire length of the U.S.–Mexico border. They document the responses of stakeholders, primarily nongovernmental organizations and U.S. and Mexican government agencies, to the border wall and propose policy alternatives.

Chapter 13

Cooperative Conservation, Unilateral Security

The Story of Two Sister Parks on the U.S.–Mexico Border

Jessica Piekielek

In a Nutshell ————————————————————

- Organ Pipe Cactus National Monument in Arizona and El Pinacate y Gran Desierto de Altar Biosphere Reserve in Sonora, Mexico, are adjacent sister parks with extensive overlapping ecological systems.

- Transboundary, cooperative management protects wildlife and habitats that span national boundaries but faces special political, logistical, and financial challenges.

- Since the late 1990s, increased U.S. border enforcement has shifted undocumented migration and drug trafficking from urban locations to wilderness areas such as Organ Pipe, where off-road driving, litter, wildlife disturbance, water pollution, soil erosion, and historical site damage result from migration, smuggling, and enforcement.

- Instead of erecting border walls, the Mexican and U.S. governments need to foster cooperation to meet the conservation challenges in transborder protected areas.

Introduction

At the heart of the Sonoran Desert sit two sister parks, Organ Pipe Cactus National Monument (Organ Pipe) and El Pinacate y Gran Desierto de Altar Biosphere Reserve (El Pinacate), where government agencies and conservation groups have nurtured modest binational conservation efforts since the 1980s. The two reserves confront dramatic changes in migration, smuggling, and U.S. border policy, resulting in environmental and social impacts. Security protocols, threats to resources, budget priorities, and changing demands on staff time place pressures on U.S. staff's capacity to work cooperatively and share resources with their counterparts in Mexico. Transboundary conservation is challenged by national politics and border security.

The U.S.–Mexico border cuts roughly in half the natural area that ecologists classify as the Sonoran Desert. Despite this ecologically arbitrary division, there exists a remarkable opportunity for transboundary conservation in the western Sonoran Desert. Both the United States and Mexico are committed to conserving this area, as demonstrated through the creation of natural reserves on both sides of the border. Close to 1.5 million hectares (3.7 million acres) of protected lands host a vast array of unusual desert and subtropical species, plus important cultural and historic sites. But rising concerns about border security in the United States overshadow and impede potential progress in transboundary cooperation, creating additional challenges for cross-border conservation.

Background

Transboundary protected areas compose 10 percent of the total protected areas worldwide, and the number of transnational conservation areas is growing.[1] Transfrontier parks and reserves are promoted as innovative mechanisms for protecting ecosystems and species that span national boundaries. The rationales for transboundary protected areas are that ecosystems and species do not conform to political boundaries and that environmental processes may link or impact multiple areas, regions, or nations. In theory, transboundary protected areas facilitate coherent management strategies that ensure ecological protection across political bounds. The authors of the edited volume *Peace Parks* go so far as to argue that transboundary conservation projects can bring participating parties together in a cooperative spirit, even when the parties may be at odds over other contentious issues.[2]

Transboundary conservation projects present several unique challenges. Transnational agreements can be politically difficult to negotiate; transnational organization presents logistical challenges; and transboundary management may be complicated where capital resources and priorities differ across borders. In addition, border areas themselves may present unique challenges to natural-resource managers. Transboundary protected areas are often located in rural, isolated settings, which can make them ideal sites for illicit transborder activities, including unauthorized migration and the illegal trafficking of people and illicit goods.[3] Kevin Dunn suggests that

creating protected areas in borderlands may further facilitate illicit transborder activities and reduce border security by maintaining the isolated nature
of these areas.[4] The question of how to reconcile the ecological theory and
intention behind transboundary protected areas with the border's political,
social, and economic realities remains a salient one.

Cooperative Conservation: Organ Pipe and El Pinacate

Organ Pipe, established in 1939 as a unit of the U.S. National Park Service
(NPS), covers 134,000 hectares (331,000 acres) of the Sonoran Desert in
southwestern Arizona. Two subsections of the Sonoran Desert, the Arizona
Uplands and the Lower Colorado River, converge within the monument,
resulting in diverse species and varied topography within the protected area.
The NPS mission is dual: to preserve natural and cultural resources and
to provide opportunities for visitor enjoyment. The monument's resource-
management program is one of the oldest continuous ecological monitoring
programs in the NPS, and staff consider Organ Pipe an important living
research center on Sonoran Desert ecology. Though off the beaten path,
the monument attracts heat-seeking winter tourists and desert lovers. Its
location on the U.S.–Mexico border was initially considered to buffer the
protected area from potential development because of northern Sonora's
remoteness and low population.[5]

Organ Pipe sits along the U.S.–Mexico border, adjacent to El Pinacate,
a federally protected area managed by the Comisión Nacional de Áreas
Naturales Protegidas (CONANP, National Commission of Protected
Natural Areas) in northwestern Sonora, Mexico. The two reserves share
about 10 kilometers (6 miles) of border. Established in 1993, El Pinacate
encompasses more than 700,000 hectares (1.7 million acres) of some of
the harshest and hottest desert in Mexico. Despite its stark nature, it hosts
several endangered species, including bighorn sheep *(Ovis canadensis)*,
Sonoran pronghorn antelope *(Antilocapra americana sonoriensis)*, and
lesser long-nosed bats *(Leptonycteris curasoe)*. As a biosphere reserve, the
area is managed in zones, with core areas that include important wildlife
habitat, scenic features, and multiuse buffer and transition zones that have
ranches and farms that predate the reserve. This management organization
matches the CONANP mission of conservation in conjunction with sustainable development and social well-being. Visitation rates at the reserve are
moderate but have increased over the years.

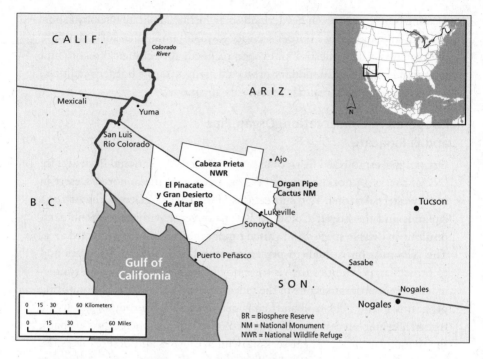

Figure 13.1. Map of Cabeza Prieta National Wildlife Refuge, El Pinacate y Gran Desierto de Altar Biosphere Reserve, and Organ Pipe Cactus National Monument. Map drawn by Mickey Reed.

El Pinacate and Organ Pipe, together with Cabeza Prieta National Wildlife Refuge, a U.S. Fish and Wildlife Service refuge directly north of El Pinacate, cover more than 1.0 million hectares (2.5 million acres) of protected lands (see fig. 13.1). This area reaches from the Gulf of California across sand dunes, dark, jagged lava fields, and delicate creosote flats to rough red mountain ranges emerging from wide desert basins. Since the 1960s, conservationists have considered various political configurations that would recognize the significance of this expanse of protected area and the ecological links between the three units.[6] Significant momentum to establish an international biosphere reserve developed in the 1980s and early 1990s, but for a variety of reasons these efforts did not come to fruition.[7]

Although the creation of an international park to include Organ Pipe and El Pinacate may be "unfinished business," there are a variety of ways that staff at the two reserves conceive of the areas as linked.[8] Visitor information provided at both protected areas points to the ecological and social links between them. Staff affectionately call Organ Pipe and El Pinacate "sister parks," and the same wording is used in materials for visitors. As one Organ Pipe staff explained, "I've always thought of Cabeza and Organ Pipe and the Pinacate as one big grand biosphere reserve, with some barriers and roads between, but functioning realistically as one big reserve, with similar plant and animal communities."[9] The parks' sister status was formalized in 2006 in an agreement between NPS and CONANP. Administrative structures and distinct missions limit the degree to which the separate units develop joint land-management plans. But staff (especially resource-management and interpretive staff) can and do easily tap into a way of thinking that brings to the fore the ecological continuity across political boundaries.

The working relationship between the two areas is a kind of *functional cooperation*, a term that Randy Tanner and his colleagues use to describe management coordination between the Glacier and Waterton peace parks on the U.S.–Canadian border.[10] Not cemented in formal political agreements, this cooperation is locally based, project oriented, and partially rooted in personal relationships between personnel at the two protected areas. Resource-management staff from El Pinacate and Organ Pipe (along with staff from other conservation agencies in the United States and Mexico) participate in several working groups aimed at transboundary conservation, including the Quitobaquito/Río Sonoyta Working Group and the Sonoran Pronghorn Recovery Team. Organ Pipe, as a better-funded and well-established park, has provided equipment, technical expertise, and staff assistance to El Pinacate as the latter develops an ecological monitoring program. In the 1990s, Organ Pipe's chief interpretation ranger started Friends of the Pinacate, a nonprofit group based in the United States (which has unfortunately folded since then). Organ Pipe staff and a local nonprofit recently applied successfully for a grant from the National Park Foundation to fund an additional staff person at El Pinacate for education and outreach. A couple of Organ Pipe staff volunteer their personal time to El Pinacate, their efforts based on personal friendships with El Pinacate staff and a fascination with the unique geography and biology of the biosphere reserve. Part of the mission of cooperation between the two protected areas has been to address the resource disparity between Organ Pipe and El Pinacate. Although both units consider themselves to be underfunded, the differences in resource access are dramatic. As the assistant director of El Pinacate pointed out to me, the Biosphere Reserve has a land mass roughly ten times that of Organ Pipe but approximately one-fifth the number of employees.

Staff at both reserves have faced challenges in developing and maintaining a cooperative relationship. Language capacity is always a concern. A handful of staff at both areas are proficient in both English and Spanish, but a lack of staff fluent in Spanish within Organ Pipe's interpretive unit limits cooperative interpretative programs in Mexico. A more official cooperative structure, rather than a link based in personal relationships, might help ensure longevity to ties between the reserves. Increases in migration, smuggling, and border enforcement in the area have more recently presented both protected areas with challenges to reserve management and cooperation.

Migration, Smuggling, and Unilateral Border Security

Since the late 1990s, remote areas of the Arizona-Sonoran border region, such as Organ Pipe, have become primary corridors for undocumented migration, drug smuggling, and border enforcement. Undocumented migration from Mexico to the United States has occurred since the early twentieth century, with public concern about illegal immigration ebbing and flowing with other political, economic, and social trends.[11] Politicians, public officials, and the media have directed much concern toward the border as a site of national vulnerability.[12] Beginning in the mid-1990s, the U.S. Border Patrol intensified enforcement in urban areas through increases in agents and construction of border walls and other enforcement infrastructure, such as stadium lights and roads. This strategy has not reduced undocumented migration but simply shifted migration routes into remote and potentially treacherous areas of Arizona.[13] Between 1995 and 2005, apprehensions in the Border Patrol's San Diego sector dropped 75 percent, but apprehensions in the Tucson sector almost doubled. Border Patrol's Ajo Station, which has jurisdiction in Organ Pipe and neighboring areas, increased its apprehensions by twenty-three times during the same time period. The Arizona-Sonora border also emerged as an important drug-trafficking corridor in the 1990s. In fiscal year 2007, the Ajo Border Patrol station seized 79,000 kilograms (174,000 pounds) of marijuana (worth roughly U.S.$174 million in the border area); the station ranked second highest for marijuana seizures among Border Patrol stations nationally. The Border Patrol has again replicated its strategy of increased agents and infrastructure development, this time in rural Arizona. The number of agents at the Ajo Station increased ten times between 1995 and 2006, from roughly twenty agents to more than two hundred. (See chapter 15 in this volume for a more in-depth discussion of the history of U.S. border policy.)

The shift of undocumented migration, smuggling, and border enforcement from urban centers into remote sections of the Arizona-Sonoran border is reshaping the social and ecological landscape. At Organ Pipe, the impacts of increased migration, smuggling, and border enforcement are multiple. First, these activities directly affect both the monument's natural and cultural resources and its wilderness value. There are an estimated 725 kilometers (450 miles) of trails and illegal roads within the monument. (See chapter 14 in this volume for a more in-depth discussion of unofficial trails and roads at Organ Pipe.) The number of people using the park to cross the U.S.–Mexico border without documentation is difficult to record. In 2005, Organ Pipe law enforcement rangers apprehended 1,040 undocumented entrants on the monument, and Border Patrol agents apprehended 25,500 undocumented entrants in the area that includes Organ Pipe. Staff cite trash, wildlife disturbance, isolated water pollution, soil erosion, changes in hydrology resulting from new roads, and damage to historical sites as some of the impacts to natural and cultural resources. Law enforcement use of the park has also increased dramatically. For example, between 1994 and 2006 the number of vehicles using one of Organ Pipe's administrative roads (not open to the public) increased by forty-four times, from 509 drive-throughs to 22,803 drive-throughs.[14] The majority of this use is likely from Border Patrol and the U.S. National Guard because the road leads to two Border Patrol camps and is an important east–west road for law enforcement access into Organ Pipe and neighboring Cabeza Prieta. Organ Pipe resource managers are still in the process of assessing the consequences of this increased activity on the natural and cultural resources that they are charged with protecting. Although they might strongly suspect, for example, that high levels of vehicle and foot traffic limit Sonoran pronghorn access to water and habitat, the scientific research required to evaluate this supposition can be complex and time-consuming.

The effects of these dramatic political and social changes go beyond direct damages to natural and cultural resources. The monument's capacity to manage its resources and provide opportunities for visitors is restricted. Citing security concerns, the park administration has closed large sections of the monument to visitor use. And segments of the park have concurrently been closed to staff or open only with an NPS law enforcement ranger escort.

These closures change with new law enforcement intelligence information, but the trend over the past several years has been toward more closures. Reduced access for monument staff frustrates their research efforts and creates maintenance and resource-management backlogs. Safety concerns and lack of access to the field affect staff morale as well.

In general, El Pinacate, with fewer staff and financial resources, faces many more obstacles in its conservation efforts than Organ Pipe does. In an unusual reversal of positions, however, Organ Pipe has witnessed many more impacts as a result of migration, smuggling, and enforcement than El Pinacate. Nevertheless, El Pinacate has not been immune. The creation of new roads and trails as well as the dumping of trash as a result of migration and smuggling are confined to a narrow stretch of the reserve between Mexico's Highway 2 (a two-lane road that crosses the northern portion of the reserve) and the U.S.–Mexico border. The portion of the reserve between the highway and border is very small and outside the most stringently protected zones. The second impact of increased smuggling involves, first, the use of dirt roads in the reserve as landing strips for drug smugglers and, second, the Mexican National Army's subsequent destruction of these makeshift airstrips. Pinacate staff have found about ten airstrips. After several years of discussion, Pinacate staff reached a negotiation with the army, which informally agreed to try more environmentally friendly techniques for deconstructing the airstrips.

El Pinacate staff act cautiously in regards to smuggling and migration through the reserve. It is not within their authority to apprehend drug smugglers or to ask to see immigration papers. The staff's policy is to report sightings of migrants to Grupos Beta (the Mexican federal task force charged with protecting Mexican migrants) and evidence of drug smuggling to the Mexican National Army. Whenever possible, staff try to avoid appearing to take sides. As one Pinacate staff member explained, "Sometimes we have come into conflict with the military. This is my personal opinion. We've come into conflict, or discussion, not conflict, discussion. Because we ask that when they destroy the airstrips that they don't destroy the vegetation around the airstrips. Sometimes they knock down saguaros or creosote bushes. And we tell them, 'You don't have to knock this down.' Then they ask us, 'Oh, so you agree with the drug smugglers?' 'No! Of course not.' But our job is to protect the environment."

Representing themselves as neutral is important for personnel safety. When I asked about safety issues, a staff member explained, "There haven't been problems because those involved in illegal activities know that our work is environmental. We aren't police. So there exists an atmosphere of, I don't want to say respect. How can I say it? We don't get involved in their affairs, and they don't get involved in environmental affairs."

Staff are quick to point out that there have been no encounters between drug smugglers and visitors or researchers in the park. Researchers, land agency staff, and so-called desert rats often described El Pinacate to me as calm and pristine compared with Organ Pipe and other border protected areas in the United States.

Challenges to Cooperation between Organ Pipe and El Pinacate

The consequences of U.S. border security policy, including shifts in migration, smuggling, and enforcement to the Arizona-Sonora border, have adversely affected·natural and cultural resources as well as management operations at Organ Pipe and El Pinacate. They have also dampened attempts by land managers and conservationists to develop and maintain strong ties between the sister protected areas. As noted previously, El Pinacate has few staff to carry out the reserve's three-pronged agenda: conservation, visitor access, and social development. Organ Pipe resource-management staff were very active in helping El Pinacate managers develop and implement its resource-monitoring and management program, including through donations of equipment and time. However, Organ Pipe personnel time has now been diverted to issues related to undocumented migration, smuggling, and border enforcement in the park. Organ Pipe staff across the resource-management, maintenance, and administrative units roughly estimate that they devote anywhere between 30 and 90 percent of their work time to border-related issues, whether in fixing the border vehicle barrier, meeting with Border Patrol, or collecting data for Organ Pipe's border impacts program. They thus have far less time to develop cooperation between the two protected areas.

In addition, after September 11, 2001, security requirements for international travel for federal employees have led to considerably less flexibility for Organ Pipe employees to travel to Mexico. Some exemptions for federal employee cross-border travel (between border parks) have been made, but only for short day trips. Although the reserves are neighbors, any overnight travel (often required for ecological monitoring work or larger infrastructure-improvement projects) to El Pinacate requires that Organ

Pipe staff get travel approval from the Washington, D.C., agency office. Processing this paperwork can take approximately five weeks, but staff may be required to submit their travel request as much as four months in advance, depending on timing and quarterly deadlines. This more stringent regulation of international travel is coupled with reduced funding and initiatives within NPS for international collaborative work in general.

Plans for construction of a border wall may further strain social and ecological ties between the two areas. In 2006, the U.S. Congress and the George W. Bush administration made various proposals for hundreds of kilometers of double fencing along the U.S.–Mexico border, including along the border between El Pinacate and Organ Pipe and Cabeza Prieta. The U.S. Department of Homeland Security built an 8-kilometer-long (5-mile-long), 4.5-meter-high (15-foot) wall along a portion of Organ Pipe's boundary, although not where it abuts El Pinacate. Nonetheless, the proposal raises the question of what it would mean to have a wall between two sister protected areas, from both an ecological and a social standpoint. When Congress passed the Secure Fence Act, which mandated 1,100 kilometers (700 miles) of double fencing on the border, most resource-management staff at Organ Pipe dismissed the proposal, suggesting that terrain and costs would prevent construction along large segments of the border. They assured their counterparts at El Pinacate that the legislation was politically significant but unlikely to be implemented, and both reserves continue to operate as though they would not see a wall between the two protected areas. El Pinacate staff, for example, have proposed wildlife underpasses for Mexican Federal Highway 2, which traverses the north end of the reserve, to improve opportunities for wildlife crossings between El Pinacate and Organ Pipe and Cabeza Prieta. These underpasses would, of course, be meaningless if a wall, impenetrable to wildlife, were constructed along the border.

Finally, a perception of Organ Pipe as a "war zone" has emerged in media and local accounts that depict the monument as unparklike. As one long-time visitor described Organ Pipe, "Once the border thing got 'hot,' there were helicopters flying up and down the highway and helicopters flying around at night, and Park Service guys in camouflage occasionally sneaking around here and there. They had operations and just the kind of things that made it unpleasant." It is a perception that Organ Pipe staff actively resist. One administrator explained, "You can still, today, go and find some very pristine wilderness areas within the park. . . . What I've really tried . . . is to change that message. . . . Yeah, we're getting impacts, we're getting lots of impacts. And there's safety issues and there's resource issues. But we still have a park worth saving here." Nonetheless, the image of Organ Pipe as chaotic and unsafe because of its proximity to the border poses challenges to land managers and conservationists working to present an alternate image of the area as the Sonoran Desert heartland, worthy of preservation. As a staff member from another U.S.–Mexico border park, Big Bend National Park, succinctly stated, "It is difficult to cooperate in resource protection when these types of 'wars' are taking place."[15]

Conclusions

The typical challenges that face many transboundary conservation efforts, such as language capacity and differential resource access, are compounded at Organ Pipe and El Pinacate by undocumented migration, drug smuggling, and an intensified and inflexible approach to border security. Organ Pipe and El Pinacate are not the only international protected areas affected. The authors of chapter 3 in this volume note similar challenges in cross-border cooperation in the Big Bend region as a result of intensified border security. And regarding Glacier-Waterton Peace Park on the U.S.–Canada border, Tanner and his colleagues discuss how rising border security concerns have directed resources away from cooperative efforts (and toward law enforcement) and restricted backcountry access between the two parks.[16]

Despite difficulties, cooperation between Organ Pipe and El Pinacate continues, primarily as a result of personal commitments and the groundwork laid by earlier binational initiatives. A political climate fostering cooperation instead of walls, more staff time, fewer travel restrictions, and increased funding would allow staff at the two parks to collaborate more effectively and extensively. Possibilities include collaborative environmental education programs in Sonoyta, Mexico, which borders Organ Pipe; more consistent joint monitoring of natural and cultural resources; and regional conservation planning to address tourism and industrial developments that may threaten both parks.

Park managers cooperate because they understand that the long-term conservation of protected areas on both sides of the border is interdependent. Transboundary ecological and social processes link the United States and Mexico. To succeed at transboundary conservation, ecological security must transcend national boundaries, and the flows of wildlife, resources, ideas, information, scientists, and conservationists will require more porous borders. But the way that we think about security in the United States is still nation bounded, as exemplified by the emphasis on walls over more synergistic responses to undocumented migration and smuggling. Successful transboundary conservation may require us to think about other issues, such as security, in a transboundary manner as well. As the authors of the articles in *Peace Parks* argue, peace parks can create opportunities for conflict resolution and reconciliation that have value even beyond conservation.[17]

Notes

1. A. Agrawal, "Adaptive management in transboundary protected areas: The Bialowieza National Park and Biosphere Reserve as a case study," *Environmental Conservation* 27(4) (2000): 326–33.

2. S. Ali, ed., *Peace Parks: Conservation and Conflict Resolution* (Cambridge, Mass.: MIT Press, 2007).

3. R. Duffy, "Peace parks: The paradox of globalization," *Geopolitics* 6(2) (2001): 1–26; M. Van Ameron, "National sovereignty and transboundary protected areas in Southern Africa," *GeoJournal* 58 (2002): 265–73.

4. K. C. Dunn, "Cross-border insecurity: National parks and human security in East Africa," in "Beyond the Arch" conference proceedings (2003), available at http://www.nps.gov/yell/naturescience/upload/proceedingsA-I.pdf, accessed June 6, 2008.

5. Southwestern Monuments Association, *Arizona's National Monuments* (Prescott, Ariz.: Prescott Courier, 1945).

6. W. L. Benner, J. Murrieta Saldivar, and J. Shepard, "Cooperation across borders: A brief history of Sonoran Desert conservation beyond boundaries," in *Dry Borders: Great Natural Reserves of the Sonoran Desert*, edited by R. Felger and B. Broyles, 560–65 (Salt Lake City: University of Utah Press, 2007).

7. C. Chester, *Conservation across Borders: Biodiversity in an Interdependent World* (Washington, D.C.: Island Press, 2006).

8. B. Siffords and C. Chester, "Bridging conservation across la frontera: An unfinished agenda for peace parks along the U.S.–Mexico divide," in Ali, ed., *Peace Parks*, 205–26.

9. Interviews with Organ Pipe and El Pinacate staff and visitors quoted in this chapter were conducted at Organ Pipe Cactus National Monument and El Pinacate y Gran Desierto de Altar Biosphere Reserve, January 25 and 29, February 14 and 15, 2007.

10. R. Tanner, W. Freimund, B. Hayden, and B. Dolan, "The Waterton-Glacier International Peace Park: Conservation amid Border Security," in Ali, ed., *Peace Parks*, 183–204.

11. K. Calavita, *Inside the State: The Bracero Program, Immigration, and the I.N.S.* (New York: Routledge, 1992).

12. J. Nevins, *Operation Gatekeeper* (New York: Routledge, 2002).

13. W. Cornelius, "Death at the border: Efficacy and unintended consequences of U.S. immigration control policy," *Population and Development Review* 27(4) (2001): 661–85.

14. Organ Pipe Cactus National Monument, *Organ Pipe Cactus National Monument Ecological Monitoring Program 2006 Annual Report* (Ajo, Ariz.: Organ Pipe Cactus National Monument, 2007).

15. J. A. Cisneros and V. J. Naylor, "Uniting la frontera: The ongoing efforts to establish a transboundary park," *Environment* 41 (1999), 12.

16. Tanner et al., "The Waterton-Glacier International Peace Park."

17. Ali, ed., *Peace Parks.*

Chapter 14

Assessing Border-Related Human Impacts at Organ Pipe Cactus National Monument

Christopher Sharp and Randy Gimblett

In a Nutshell

- Beginning in the late 1990s, impacts to protected lands along Arizona's border with Mexico began increasing due to border-enforcement strategies employed elsewhere. These impacts have been documented and studied at Organ Pipe Cactus National Monument in south-central Arizona.

- The various users of protected land along the border (undocumented migrants, drug runners, law enforcement personnel, citizen watch groups, humanitarian aid organizations, recreationists, land managers) have characteristic patterns of use that can be quantified.

- These impacts have increased and are cumulative and interrelated in scope and scale.

- Because quantified significant impacts—on designated Wilderness, biological diversity, hydrological function, and other natural assets—are in direct conflict with original protection mandates, difficult questions arise for land managers about acceptable limits of change.

- Some natural-resource-management solutions will rely on better coordination between law enforcement and land-management agencies.

- Decisions about future management need to address the interrelated impacts of enforcement and undesirable activities.

Introduction

Lands adjacent to the U.S.–Mexico border are undergoing dramatic changes in their social and environmental history. Undocumented migration, especially the most recent wave of people entering the United States through Mexico, is a complex topic that has dominated regional- and national-level political discussions for the past decade or more. The biophysical impacts from unauthorized traffic through Organ Pipe Cactus National Monument

(Organ Pipe), which shares 48 kilometers (30 miles) of the international border between Arizona and Sonora, demonstrate that public lands are at the center of this issue (see figure 13.1 in chapter 13 of this volume).

In recent years, Organ Pipe and other public lands have become increasingly popular locations for drug smugglers and undocumented migrants crossing from Mexico into the United States. (In biological terms, "migration" is a species' periodic, seasonal travel along a well-established route. In this study, the term *migrant* is used more narrowly to denote people traveling in only one observed direction, into the United States.) In the mid-1990s, U.S. Border Patrol and U.S. Customs increased enforcement activities in more accessible urban areas of Arizona (e.g., Douglas and Nogales). These efforts ultimately diverted large numbers of undocumented migrants and drug smugglers into more remote areas, including several national park units. A dramatic increase in resource impacts has resulted from illegal entry, law enforcement, and homeland-security initiatives along the U.S.–Mexico border. These impacts threaten the unique flora and fauna of transboundary ecosystems such as the Sonoran Desert. In addition, they have posed a considerable challenge to land stewards charged with developing conservation strategies and managing these unique and sensitive areas. Although evidence of impacts at Organ Pipe exists, little work has been done to determine if a consistent and predictable pattern of use and impacts can be established.

This chapter presents observations of human impacts at Organ Pipe between 2004 and 2007, and provides insights that may be applicable to other border protected lands facing similar impacts. It begins with a background discussion of previous research regarding human impacts on protected areas and a description of the study area. It then reports on primary results from the authors' study at Organ Pipe. It concludes with a broad discussion of the implications of growing human impacts on conservation and protection of natural resources on this and other public lands.

Background

Human Impacts in Protected Areas

In resource-management terms, *visitor impact* denotes "any undesirable visitor-related biophysical change of the wilderness resource."[1] Previous research on recreation has established that visitors to protected areas impact four primary ecosystem components: soil, vegetation, wildlife, and water.[2] Impacts on a single ecological element can have cascading effects on the ecosystem. For instance, user-created trails or campsites may cause landscape fragmentation, potentially interfering with movement of some animal species. Localized disturbances can disrupt wildlife, damage sensitive habitats, and cause erosion. Stream sedimentation from trail and campsite erosion can reduce the quality of aquatic habitats for insect and fish populations. In addition, the presence of visitors may displace wildlife from essential habitat. Some studies have also shown that impacts perceived by subsequent visitors degrade the quality of their experience. Critically, the biophysical impacts of human uses are not evenly dispersed across landscapes. They can occur very quickly with even low levels of use, and recovery can be very slow in ecosystems such as the Sonoran Desert.

For managers of protected areas, problems have long existed in establishing standards for acceptable impact levels and for appropriate activities within particular settings.[3] For managers along the U.S.–Mexico border, these problems are coupled with a growing concern about how best to understand and minimize impacts associated with illegal or unauthorized uses.

Site History of Organ Pipe Cactus National Monument

Protected areas such as Organ Pipe, Coronado National Memorial, Coronado National Forest, and Buenos Aires and Cabeza Prieta National Wildlife Refuges compose the majority of land along the U.S.–Mexico border in Arizona. Located near the geographic center of the Sonoran Desert, Organ Pipe shelters a rich complement of flora and fauna. Several species, including the organ pipe cactus *(Stenocereus thurberi)* and senita cactus *(Pachycereus schottii)*, reach their northern distributional limit within the monument. The United Nations Educational, Scientific, and Cultural Organization designated Organ Pipe an International Biosphere Reserve in 1976 based on its significance as a representative biological region of the world. Approximately 94 percent (126,600 out of 133,600 hectares [312,600 out of 330,000 acres]) of this remote monument was also federally designated as Wilderness in 1978 based on its untrammeled character and the presence of four federally listed endangered species: the Sonoran pronghorn *(Antilocapra Americana sonoriensis)*, the lesser long-nosed bat *(Leptonycteris curasoae)*, the Quitobaquito pupfish *(Cyprinodon eremus)*, and the Mexican Spotted Owl *(Strix occidentalis lucida)*.[4]

The U.S. Border Patrol estimated in 2002 that an average of 300 to 500 people per day entered the United States illegally through Organ Pipe.[5] However, according to area nongovernmental agencies, this estimate may represent less than half of the actual total. Combined with park visitation, which exceeds 295,000 individuals per year, this significant, additional use represents the potential for an exponential increase in impacts to park resources.[6]

In past years, the barbed-wire fence marking the international border along the southern boundary of Organ Pipe failed to prevent people from illegally entering the United States there. In 2006, the U.S. National Park Service constructed vehicle barriers across Organ Pipe's international border. Although the vehicle barriers significantly reduced off-road vehicle traffic in many areas, they could not prevent cross-border foot traffic.

Networks of foot trails have formed in a pattern running from south to north throughout Organ Pipe. Fires lit by undocumented migrants for warmth or as rescue beacons sometimes result in wildfires. Large volumes of abandoned personal belongings and debris accumulate along travel routes, at overnight sites, and around water sources. Some individuals paint or inscribe graffiti on rocks and cactus. Direct impacts to resources from these activities include erosion, vegetation damage and complete vegetation loss in some areas, water contamination, wildlife disturbance and mortality, archaeological resource damage, and monument infrastructure damage. Indirect impacts can include changes in overall surface hydrology, groundwater depletion, and habituation of wildlife to trash dumps.[7]

This situation has posed considerable challenges to Organ Pipe's land stewards. In response, in 2004 the monument launched an interdisciplinary effort to develop inventory and monitoring processes to define border impacts qualitatively. This team is also developing methods to manage and mitigate the associated operational and environmental impacts. In 2005, the monument developed a five-year strategic plan and an official resource-management goal to address border impacts. Both plan and goal identify the need to ensure that the monument's natural and cultural resources are protected and restored through science-based decision-making processes. In effect, Organ Pipe has refocused its entire resource-management program to address border impacts.

Study of Direct Impacts and Patterns of Human Use in Organ Pipe Cactus National Monument

Methods

Although several studies document human impacts in Organ Pipe, no previous study has set out to identify and understand if a consistent and predictable pattern of use and impacts exists. Our study began in 2004 to examine the observable trail proliferation within the monument's boundary, using satellite imagery, ground verification, and interviews of local professionals. We recorded and georeferenced observable linear impacts (trails) throughout the monument. Then we compared the satellite imagery, ground data, and interviews to evaluate the accuracy of the methods. The study used four different image sets with dates ranging from 2002 to 2007 to evaluate changes in impact. We statistically evaluated the final geodatabase for trend analysis and pattern recognition. This study's quantitative assessment is the first step in integrating user impacts and the behavior patterns associated with distinct groups of users. Next, we present results from the georeferenced data and a description of user groups.

Change Detection

In the face of greater cumulative legal and illegal uses of Organ Pipe resources, the question arises, "How much change in natural conditions has occurred, and is it acceptable?" To examine the changes, a comparative study of travel routes was conducted using four sets of georeferenced satellite imagery over six years, which allowed for a greater probability of detection and an understanding of changes in trail proliferation over time.

In 2004 and 2005, recognizing an immediate need to document the extent of illegal off-road vehicle tracks and foot trails occurring in Organ Pipe's wilderness lands, monument staff developed a series of protocols to map the distribution of and quickly assess impacts to monument resources from these activities. These protocols included a survey of the monument's administrative and public-use roads, documenting roadside damage and off-road driving incursions. Mapping off-road vehicle routes with the use of global positioning system equipment documented the actual routes of travel. A "rapid assessment tool" was used to map and provide a qualitative assessment of impacts along foot trails and some off-road vehicle routes. Satellite imagery supplemented the ground-based mapping effort. The right side of figure 14.1 depicts the routes that have been detected, along which illegal and law enforcement activities occur within Organ Pipe. It is evident that an extensive trail system has developed throughout the monument, distinct from the trails that are planned and maintained by its managers, as seen more clearly on the left side of figure 14.1.

From 2004 to 2005, by physically walking and mapping individual routes, monument staff documented more than 579 kilometers (km) (360 miles [mi]) of off-road vehicle routes in Organ Pipe.[8] Approximately two-thirds (385 km [239 mi]) of these routes were created before 2004, and

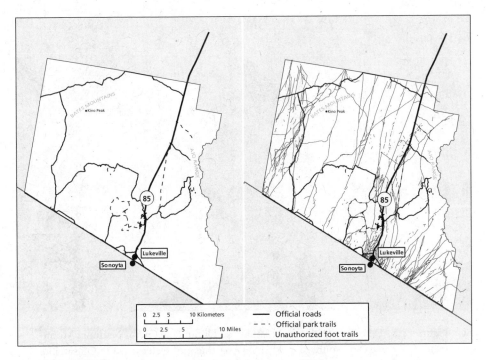

Figure 14.1. *Left:* Official roads and recreation trails within Organ Pipe Cactus National Monument boundaries. *Right:* Official routes overlain with unauthorized routes throughout Organ Pipe.

one-third (194 km [121 mi]) were created during the observation period. Routes varied from single- to multiple-use tracks.[9] In 2006, by physically walking and mapping individual trails, monument staff documented more than 500 km (311 mi) of illegal foot trails in Organ Pipe.[10] In addition, east–west monitoring transects, covering a total of 117 km (73 mi) in the south, central, and north portions of the monument captured 452 foot trails (averaging 3.9 per km), 85 vehicle tracks (0.7 per km), 10 more-established roads (0.1 per km) traveling in a north–south direction, and 386 pieces of trash (3.3 per km). Route frequencies were greater for transects located closer to the U.S.–Mexico border (5.6 per km) than for interior (4.0 per km) or northern (4.4 per km) transects, suggesting a more dispersed pattern of travel by people upon crossing the U.S.–Mexico border as compared with their travel in the monument's interior.[11]

Figure 14.2 illustrates the patterns of user groups over time. U.S. Border Patrol agents, undocumented migrants, and smugglers use this area heavily,

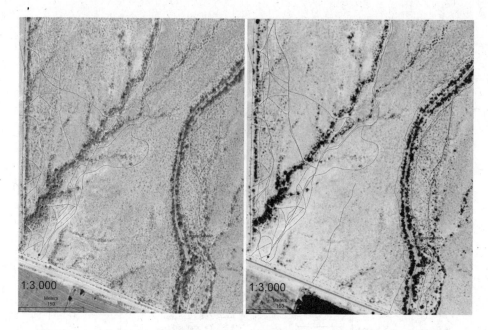

Figure 14.2. Detection of change in impact on Organ Pipe Cactus National Monument over time. Data sources: *(left)* U.S. Department of Defense IKONOS satellite image, October 25, 2003; *(right)* U.S. Department of Agriculture National Agriculture Imagery Program images, June 24, 2007, to July 1, 2007.

causing significant impact. The image shows an area east of Lukeville, Arizona, in Organ Pipe. The U.S.–Mexico border is on the southern (bottom) edge of the photo, and dirt roads allow access along both sides of a vehicle barrier. In the older image (fig. 14.2, left), significant vehicle impact appears on the western edge. This vehicle impact can be linked to Border Patrol enforcement activities.[12] This route was used to clear tracks from open ground and trail crossings to count passersby and access by illegal users. From the image alone, it is not possible to determine which user group initiated the tracks in this area, only to estimate the total level of use. The trails along the eastern edge of the image fit the pattern of illegal users, conforming to higher-concealment areas along the riparian zone. Although riparian areas are more susceptible to impacts, this user type rarely leaves a trail wider than a single track, which is harder to see.

The image on the right side of figure 14.2 clearly reveals impact change over time. The western section shows more open ground and diminished vegetation (dark images in the figure). A larger number of vehicle-width tracks can also be seen inside of the original impact area. The pattern on the eastern (right) edge of the image demonstrates an increase in impact as well, with a proliferation of trails functionally duplicating existing trails. These trails continue to be narrow and difficult to differentiate from the landscape. Without addressing which type of impact is more detrimental to the ecosystem, one can still clearly differentiate the patterns of impact.

There are numerous regional maps of routes across the border region: historic routes marked by the writings of Padre Eusebio Francisco Kino and Juan Bautista de Anza, the traditional routes of the Hohokam and Pápago/Tohono O'odham known and marked by use, and more modern trade and wagon routes. All predate current border traffic patterns. The impact patterns mapped over Organ Pipe show a distinct pattern of use across the landscape. Zooming in on the corridors shown in figure 14.1 would illustrate significant variability, but large-scale patterns of use can clearly be seen without that. The routes at the western (left) part of the image closely conform to the Hohokam shell trail, which in turn is defined by the Bates Mountains and Kino Peak to the east. Other concentrations in the center show the influence of U.S. Highway 85 as a navigational aid and pick-up/drop-off route. Eastern (right-hand) patterns coalesce along mountain passes in the Ajo Range leading out of Organ Pipe and into the Tohono O'odham Reservation. Natural features such as mountain ranges and water sources as well as human installations such as roads and communities seem to drive metascale human impact patterns.

Usage Vectors

As described earlier, direct impacts from park users can result in a mosaic of change to an ecosystem, followed by natural responses to that impact and the subsequent feedback process as other users respond to that impact (see fig. 14.3). Multiple vectors—from recreation users (both rule abiding and non–rule abiding) to illegal users (including undocumented migrants and drug smugglers) and official management actions—represent very different patterns of impact at Organ Pipe (and in lands throughout the border region). These distinct user groups are the logical explanation for unique and spatially distinct impact patterns observed in the geodatabase. Groups such as humanitarian aid organizations and civilian watch groups were also included in the study but are currently considered a less-significant source of impacts. Next, we provide greater detail about each usage-type vector common to borderland protected areas, which aids in our interpretation of the physical impact data at Organ Pipe.

Figure 14.3. Examples of human impact in Organ Pipe Cactus National Monument. *Clockwise from the top left:* an area of impacts caused by frequent human use and erosion; an area impacted by illegal usage where vegetation has been removed and trash has accumulated; a single track route likely used for drug smuggling; a staging area for the U.S. Border Patrol that suffers from severe vegetation loss and soil compaction.

Recreation Users

Recreation users of protected areas are known to pursue the individual goals of solitude, aesthetically pleasing landscapes, and adventure.[13] Recreation managers use a system of signage, access controls, and maintenance to help limit impact from legitimate recreation use. Resource managers also regulate access to minimize the potential for user conflicts. Conflict occurs when recreation users encounter interference with their goals—for example, through negative aesthetic impacts to the landscape or encounters with undesired users. Users may then be displaced and seek recreation experiences elsewhere or at other times, resulting in a change to the patterns of resource use and impacts.[14]

An assessment of the sources of impacts to natural areas must also address antisocial or unsustainable recreation types. Some users of public lands seek out isolated areas as the setting for dangerous, illegal, or unsustainable activities. These activities include trail or road cutting by off-road vehicles, out-of-control parties, and destructive recreational shooting of improvised targets. These users' behavior is marked by a lack of awareness of the impact or conflict their activities create. Managers typically try to limit these impacts with a combination of education and enforcement. Prohibited activities are typically displaced by enforcement rather than eliminated; in other words, they are moved to less-regulated or enforced areas.

Undocumented Migrants

Undocumented migrants represent a complex group of borderland users that has a significant impact on wilderness lands. Individual, unaffiliated migrants exist, as do migrants who are connected to organized operations that move throughout the border area. Although both types of users have the common goal of crossing the border area undetected, their behaviors differ significantly.

Unaffiliated undocumented migrants traveling singly or in small groups lack specific local knowledge of law enforcement, terrain, and resources. These individuals can behave in apparently random ways, sometimes damaging resources and endangering themselves in order to avoid detection.[15] These behaviors lead to conflicts with law enforcement officers and a fear of the natural environment. Helpful landscape features such as existing trails, water sources, and humanitarian aid stations are viewed as potential traps and thus avoided. Unaffiliated migrants' behavior is marked by dispersion of use and a deep lack of awareness of the region's natural resources.

Organized undocumented migrants have a similar goal—to cross undetected—but so long as the organization remains intact, they behave differently. The guides (known as *coyotes*) of these groups, ranging from three to twenty-five people, typically travel only within a known territory, relying on their familiarity with the route to avoid detection. The natural resource is seen as a component of avoiding detection. Viewing the natural resources as useful rather than an antagonistic element dramatically changes the organized migrants' behavior. Specific trails are preferred, chosen for characteristics that hinder detection—such as speed of travel, avoidance of known hazards, and resources such as water and shelter.[16] Groups are sometimes abandoned by the coyotes, and individuals are separated or criminally preyed upon. In these situations, it is reasonable to expect that people will behave like unaffiliated migrants, with uncertainty and distrust.

Drug Smugglers

Although one method of drug smuggling is to intermingle with groups of undocumented migrants, drug smugglers represent a distinct user group. The high dollar value of drug loads leads to a predictable increased effort to ensure success crossing the border. Smugglers accordingly employ strategies such as sophisticated lookout posts, decoy groups, threats, and intimidation. Their use of natural resources to conceal smuggling activities exists at a higher level than with other users. An effort is made to hide the record of their passage and actively to avoid encounters with law enforcement and others during the crossing. This effort appears to be an attempt to maintain clandestine travel routes and to conceal the volume of actual traffic through the borderlands. Law enforcement teams identify drug trails by the characteristic use of slick rock or difficult terrain to avoid detection and by the practice of leaving only narrow tracks to obscure numbers of individuals (see fig. 14.3, bottom right).

Resource Managers and Law Enforcement Agencies

The impacts of each of these groups on local ecosystems and the larger-scale social implications of illegal use of border public lands inspire a significant response from natural-resource managers and law enforcement agencies. An important feedback relationship between illegal uses and law enforcement responses also causes damage, however. The impacts of both groups are interrelated, although the two groups' goals are dramatically different. The impacts produced by one group are expected to shift as the other's intensity or methods shift.

Understanding management groups' priorities helps to identify the impact patterns that their actions produce. Numerous agencies are tasked with law enforcement in the borderlands of southern Arizona, including the U.S. Border Patrol, U.S. National Park Service, U.S. Forest Service, state and federal fish and wildlife agencies, and state, county, and tribal police. Although agency mandates defining methodology and jurisdiction for these groups vary, their collective goals are public safety and deterrence of illegal activity. Because law enforcement is by nature responsive, these agencies are typically forced to react to the actions and changes of illegal user groups. Responsive tactics, focused on speed and safety, often lead to off-road vehicle use by officers. As a result of this type of activity, large tracts of desert are impacted (see fig. 14.3, bottom left).

Next Steps

This study effectively captured patterns of user impacts using satellite imagery within Organ Pipe's boundary. However, some of the questions about user type and causal relationships in impacts would require real-time observations of users. The borderland user groups described in this chapter vary in their interactions with the landscape. Each user group has distinct goals that shape those interactions. For researchers studying user impacts, each user group's relationship to the landscape also shapes the expected patterns of impact connected to the user. A robust, unbiased sample of Organ Pipe visitors is a practical impossibility given legal, safety, and technological constraints. An effort to integrate this data with user-type data derived from trail observation, trace data, or other means would add significantly to understanding the patterns observed.

Conclusions: Implications for Future Conservation, Protection, and Management of Lands along the U.S.–Mexican Border

Although the political boundaries and the rationale for travel across the border have changed over time, underlying factors remain constant. This region's natural resources have a limited ability to absorb and recover from the impacts of human use. Human impacts continue to proliferate across protected areas such as Organ Pipe Cactus National Monument. Humans traveling across the border region illegally try to avoid detection, creating a dispersed pattern of impacts. These impacts range from degradation and loss of desert plant communities to large denuded areas with high degrees of soil compaction and susceptibility to erosion. Disturbance to threatened and endangered wildlife and higher mortality rates due to increased vehicle traffic remain prevalent. Garbage and destruction of archaeological resources are significant concerns for land-management agencies. Of grave concern is the proliferation of trails and travel routes, which help to disperse these primary and secondary impacts across the landscape. Such impacts clearly have increased at Organ Pipe in a very short time period, leading to a more dispersed and greater overall ecosystem impact that has cascading and devastating effects across the landscape. Sensitive, semiarid environments such as the Sonoran Desert have been observed to recover from such impacts extremely slowly. Little is known about whether the human-caused impacts already observed actually exceed the ecosystem's ability to recover. Only time and long-term monitoring can provide a glimpse of these recovery patterns.

At the time of this writing, many sections of Organ Pipe are closed to the public due to security and safety concerns associated with illegal uses, and it appears that they will remain closed indefinitely. Although protection of the public is a justification for these closures, resource impacts continue to accelerate due to the illegal use and law enforcement response. The public is largely unaware of the impacts from these interactions and know only that large tracts of the resource are off-limits. In an area formally designated as Wilderness, set aside for its primitive wilderness character, unconfined use notably seems to be in conflict with today's land-management imperatives. The notion of using a landscape and leaving no imprint as written in the Wilderness Act is far from realistic at Organ Pipe under current conditions.[17]

Not only has there been an increase in impacts along the border from illegal activity, but the presence of additional law enforcement under U.S. Department of Homeland Security initiatives has significantly added to these impacts. The combined impacts are prolific and continue to escalate as greater resources outside of the local land-management agencies' control are added to the border region. Although enforcement is needed along the border, greater awareness and understanding of the extensive impacts that can occur as a result are also required.

Management of protected areas occurs within highly politicized contexts where the objective of preserving natural areas is frequently affected by the desire to encourage recreational use. The decisions about how to manage legal, illegal, and law enforcement activities start with understanding the movement patterns of human traffic and the impacts associated with these patterns, and then move on to defining realistic goals from which to develop management prescriptions. Protected areas along the U.S.–Mexico border face immense pressure and threats to their ecological integrity. In addition, wilderness values are eroded by the influx of illegal and law enforcement activities. Organ Pipe Cactus National Monument is just one of many protected areas feeling this immense pressure. The time has come to ask for what values are these areas to be managed. This political and value-laden decision needs to be addressed before a unique and biologically diverse portion of the Sonoran Desert is destroyed beyond repair. The work needed to prevent this outcome starts with effective cooperation between land-management and law enforcement agencies as they work toward a common set of goals through creative solutions to achieve natural-resource protection in the context of intensified human impacts in borderland protected areas.

Notes

1. Y. Leung and J. Marion, "Recreation impacts and management in wilderness: A state-of-knowledge review," in *Wilderness Science in a Time of Change Conference*, compiled by D. N. Cole, S. F. McCool, W. T. Borrie, and J. O'Loughlin, vol. 5 of *Wilderness Ecosystems, Threats, and Management*, Proceedings RMRS-P-15-VOL-5 (Ogden, Utah: U.S. Department of Agriculture, Forest Service, Rocky Mountain Research Station, 2000), 23.

2. Ibid.

3. D. N. Cole, "Monitoring and management of recreation in protected areas: The contributions and limitations of science," in *Policies, Methods, and Tools for Visitor Management: Proceedings of the Second International Conference on Monitoring and Management of Visitor Flows in Recreational and Protected Areas, June 16–20, 2004, Rovaniemi, Finland, 9–16* (Saarijärvi, Finland: Gummerus, 2004).

4. Wilderness Act, Public Law 88-577, 88th Congress, 4th sess., September 3, 1964.

5. U.S. General Accountability Office (GAO), *Border Security: Agencies Need to Better Coordinate Their Strategies and Operations on Federal Lands*, GAO 04-590 (Washington, D.C.: U.S. GAO, June 2004).

6. Ibid.

7. Leung and Marion, "Recreation impacts and management in wilderness."

8. S. Rutman, *A Foundation for Collecting, Managing, and Integrating Information on the U.S./Mexico International Border Activities, June 2004–September 2005* (Ajo, Ariz.: Organ Pipe Cactus National Monument, U.S. National Park Service, 2006).

9. Ibid.

10. Organ Pipe Cactus National Monument, unpublished data.

11. A. Povilitis and E. Fallon, *East–West Transect Report for 2006, Organ Pipe Cactus National-Park* (Washington, D.C.: U.S. National Park Service, 2006).

12. Tim Tibbitts, law enforcement officer, interviewed by Christopher Sharp, October 5, 2007, Organ Pipe Cactus National Monument.

13. R. E. Manning, *Studies in Outdoor Recreation* (Corvallis: Oregon State University Press, 1999).

14. D. H. Anderson and P. J. Brown, "The displacement process in recreation," *Journal of Leisure Research* 16(1) (1984): 61–73.

15. J. Annerino, *Dead in Their Tracks: Crossing America's Desert Borderlands* (New York: Four Walls Eight Windows, 1999; Tucson: University of Arizona Press, 2008).

16. Leung and Marion, "Recreation impacts and management in wilderness."

17. Wilderness Act, Public Law 88-577.

Chapter 15

A Fence Runs Through It

Conservation Implications of Recent U.S. Border Security Legislation

Brian P. Segee and Ana Córdova

In a Nutshell ————————————————

- The 2005 REAL ID Act gave the secretary of the U.S. Department of Homeland Security authority to waive laws as necessary to hasten border wall and road construction.
- In 2006, the U.S. Congress rejected a comprehensive overhaul of immigration policy and instead passed the Secure Fence Act, which required the construction of 1,129 kilometers (700 miles) of walls and barriers along the border.
- In combination, these laws have facilitated extensive border wall construction without prior environmental analysis, resulting in ecological degradation and creating new environmental conflicts in the border region.
- Comprehensive immigration reform is needed, along with a repeal of the REAL ID waivers.
- Interim environmental legislation must provide for better coordination, study, and mitigation efforts related to current activities, including binational cooperation with Mexican institutions.
- Civil society, nongovernmental organizations, and scientists must continue to demonstrate that the shared border environment is being negatively affected.

Introduction

Under recent legislation known as the Secure Fence Act of 2006, the U.S. Congress directed that the Department of Homeland Security (DHS) construct more than 1,129 kilometers (700 miles) of border wall along the southwestern border, including numerous areas of protected federal lands, habitat for threatened and endangered species, and other natural resources.[1] Although there is obvious potential for an infrastructure project of this magnitude to have significant environmental consequences, relatively little data address the border wall's predicted impacts. This dearth of information has been exacerbated by a second law passed by Congress, the REAL ID Act, which grants the DHS secretary power to waive all laws she or he deems necessary to ensure the expeditious construction of the border wall and associated roads. Between 2005 and 2008, former DHS secretary Michael Chertoff invoked this authority on five occasions, including one decision waiving the application of thirty-five laws to more than 757 kilometers (470 miles) of border wall construction. As a result, DHS did not conduct environmental analysis under the National Environmental Policy Act (NEPA), Endangered Species Act, and other laws normally applicable.

Under these waivers, sections of the border wall have already been completed in several areas of protected federal lands, including national wildlife refuges, monuments, and wilderness areas. Despite the relative lack of information, biologists with the U.S. Fish and Wildlife Service (FWS) and other experts believe that completed and proposed border wall construction will have numerous environmental consequences, including blocking transboundary movements of imperiled wildlife such as the jaguar *(Panthera onca)* and ocelot *(Leopardus pardalis)*.

In order to better understand current U.S. policy and identify alternatives, this chapter chronicles recent actions by the U.S. Congress in relation to the border wall and their implications for the environment and addresses several questions: How and why did Congress choose to mandate extensive border wall construction, while exempting such construction from compliance with federal laws, much of it within protected federal lands and essential wildlife habitat? Where has construction already occurred, and what will be the likely environmental consequences? How have various stakeholders responded to border wall construction? In addition, we provide examples of transboundary conservation initiatives that are threatened by border security efforts. Finally, we examine alternatives to and provide recommendations for better integrating U.S. border security law and policy with environmental considerations.

Background and Discussion

U.S. Congress and the Border Wall

In 2006, the U.S. Congress, with the support of President George W. Bush, had an opportunity to pass the most significant reform of U.S. immigration law in two decades. Although the House of Representatives had instead passed an "enforcement-only" bill in December 2005 that would have required 1,129 kilometers (700 miles) of border wall construction and made it a felony for undocumented immigrants to remain in the country (H.R. 4437), that bill generated widespread public opposition and large protests in several cities. The Senate then passed a bipartisan, broadly supported, comprehensive immigration bill that mandated 483 kilometers (300 miles) of border wall construction and strengthened workplace-enforcement provisions, but also included a revamped framework for legal immigration and would have provided a path to citizenship for most undocumented immigrants (S. 2611). Under well-organized pressure from dissenting members of the House who refused even to consider the Senate's citizenship provisions, however, House and Senate leaders never convened a conference committee to resolve differences between their two bills.[2]

After the collapse of comprehensive reform efforts, many members of Congress remained motivated to pass at least one piece of immigration legislation before the midterm elections. In the fall of 2006, the House introduced and overwhelmingly passed 283 to 138 the Secure Fence Act of 2006 (H.R. 6061). The Senate passed the bill two weeks later without amendment, also by an overwhelming 80 to 19 majority.[3]

Less than three pages long, the Secure Fence Act is silent on the contentious immigration policy issues that had divided the House and Senate, instead focusing solely on border wall construction. Under the act, DHS was required to "achieve operational control" of the border within eighteen months and is specifically directed to erect five segments of border wall, additional barriers, roads, lighting, cameras, and sensors on approximately 1,129 kilometers (700 miles) of border in California, Arizona, New Mexico, and Texas. Despite the fact that these 700 miles encompass extensive areas of protected federal lands, habitat for threatened and endangered species, and other sensitive natural resources, the potential environmental consequences of border wall construction were scarcely considered during Congress's brief debate on the legislation. The Secure Fence Act was subsequently amended to remove requirements that specific sections of wall be built, but still directs DHS to construct "not less than 700 miles" of border wall.

In addition to legislating border wall construction as a response to immigration concerns, Congress has also cast wall construction as a component of antiterrorism policy. In 2005, it had also enacted a sweeping provision permitting the DHS secretary to waive laws that would otherwise apply to border wall construction. Under section 102 of the REAL ID Act (H.R. 418), a bill largely directed at antiterrorist efforts in the wake of the September 11,

2001, attacks, the DHS secretary was given the "sole discretion" to waive all laws she or he "determines necessary to ensure expeditious construction of the barriers and roads" along and "in the vicinity" of all 9,656 kilometers (6,000 miles) of U.S. international borders with Mexico and Canada—a remarkable grant of powers to a politically appointed official.[4] Yet the REAL ID Act was passed as a rider to so-called must-pass legislation—an emergency bill that provided funds for the wars in Iraq and Afghanistan as well as tsunami-relief efforts—and thus was never given committee consideration or hearings.[5]

Walls without Laws: The San Pedro Riparian National Conservation Area

The San Pedro Riparian National Conservation Area (San Pedro NCA), a 22,662-hectare (56,000-acre) oasis of verdant riparian and grassland habitat in southeastern Arizona set aside by Congress in 1988 to protect the San Pedro River, vividly illustrates the interdependence of U.S.–Mexico transboundary ecosystems and the manner in which the Secure Fence Act and the REAL ID Act waiver authority have acted together to speed the construction of border wall segments without compliance with key environmental laws. Indeed, the San Pedro River, one of the last free-flowing rivers in the desert Southwest, is globally recognized for its avian and mammalian diversity and serves as an important conduit for wildlife movement between the United States and Mexico. Congress recognized these values in its designation of the San Pedro NCA, directing the U.S. Bureau of Land Management (BLM) to protect the river's fragile riparian habitat and wildlife, and to preserve the area's archaeological, scientific, cultural, and recreational values.

The Secure Fence Act, however, originally called for border wall construction along most of the Arizona border, including the southern limit of the San Pedro NCA, and in October 2007 private contractors working on behalf of DHS began constructing a wall and new road within its boundaries. Although the U.S. government had built other wall segments during the previous decade, construction had been sporadic and largely limited to heavily populated urban areas such as San Diego and El Paso. DHS's building of the border wall within the San Pedro NCA was the first time such construction had occurred on public land specifically protected for its environmental resources. The BLM, despite its legal mandate to protect the area's natural and cultural resources, had approved a cursory environmental analysis prepared under NEPA and granted DHS a perpetual right-of-way for the construction without public notification or involvement.

Conservation organizations Defenders of Wildlife and the Sierra Club believed that the BLM's decision threatened to establish a precedent for future border wall proposals within the rich tapestry of public lands along the U.S. side of the U.S.–Mexico border and that the proposed construction would result in certain environmental consequences within this unique area. They filed a federal lawsuit against BLM and DHS, and then obtained a temporary injunction against further construction. A federal judge agreed that the BLM had likely violated NEPA and other laws in its approval of the construction and that irreversible harm to wildlife and the San Pedro River's riparian habitat would occur absent such injunction.

The conservation groups sought to bring DHS to the negotiating table over security alternatives to wall construction within the San Pedro NCA. Instead, DHS secretary Chertoff in October 2007 invoked his sweeping powers under the REAL ID Act to waive the application of nineteen laws, including the Endangered Species Act, the National Historic Preservation Act, the Archaeological Resources Protection Act, and the Safe Drinking Water Act.

DHS then moved quickly to complete the border wall construction on San Pedro NCA, the first such action on federal protected lands relying on the waivers, although Secretary Chertoff had already used the waiver to expedite construction on other wall segments in San Diego and on the Barry M. Goldwater Air Force Range in southwestern Arizona. On April 1, 2008, Chertoff again invoked REAL ID to waive thirty-five laws—including NEPA and the Endangered Species Act as well as laws intended to protect archaeological and cultural resources, public health and safety, and Native American religious and burial sites—that would otherwise have applied to more than 756 kilometers (470 miles) of proposed wall construction in all four southwestern border states (see figs. 15.1 and 15.2).

Environmental Impacts of Border Wall Construction

DHS has also invoked its authority under the Secure Fence and REAL ID acts to build border walls and roads within numerous other areas of protected federal lands, including the Otay Mesa Wilderness Area (California), Buenos Aires National Wildlife Refuge (Arizona), Coronado National Memorial (Arizona), and Lower Rio Grande National Wildlife Refuge (Texas) (figs. 15.1 and 15.2). Additional wall construction is proposed on more federal

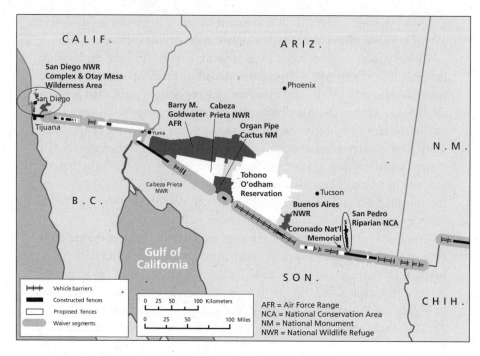

Figure 15.1. Western portion of U.S.–Mexico border, barriers, and waiver segments as of December 2008. Data on segments compiled by Matt Clark.

lands and protected private lands, including the Audubon Society's Sabal Palm Audubon Center and Sanctuary (Texas) and The Nature Conservancy's Lennox Southmost Preserve (Texas).

Completed and proposed wall construction on protected lands and other areas promises significant environmental consequences. Although data are limited—a fact exacerbated by DHS's sweeping REAL ID waivers and associated failure to conduct meaningful environmental analysis— numerous impacts are anticipated, and many have already occurred. FWS, for example, identifies border wall impacts as including ground disturbance, vegetation removal, soil compaction, and hydrological disruption and erosion. FWS also believes that border wall construction may displace wildlife due to disturbance from human presence, noise, and artificial light, and in some instances will result in direct wildlife mortality.[6]

Border walls and roads act as a barrier to wildlife movement, posing risks to species that rely on habitat within both the United States and Mexico, including jaguar (*Panthera onca*), black bear (*Ursus americanus*), Sonoran

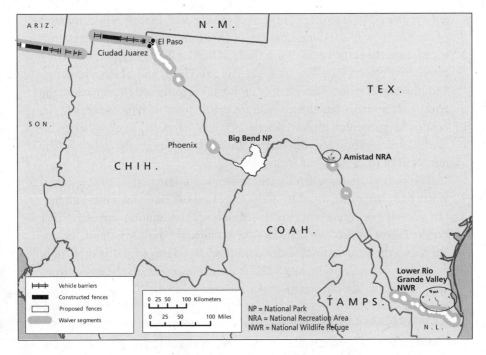

Figure 15.2. Eastern portion of U.S.–Mexico border, barriers, and waiver segments as of December 2008. Data on segments compiled by Matt Clark.

pronghorn antelope *(Antilocapra americana sonoriensis)*, Mexican gray wolf *(Canis lupus baileyi)*, bison *(Bison bison)*, ocelot *(Leopardus pardalis)*, and Cactus Ferruginous Pygmy-Owl *(Glaucidium brasilianum)*.[7] In some cases, wall construction may result in extirpation of rare wildlife from the United States; for example, biologists believe that extensive wall construction in Arizona will likely preclude jaguar recovery and establishment of breeding populations within the United States.[8] Similarly, FWS, which manages the Lower Rio Grande National Wildlife Refuge, has predicted that wall construction will isolate and limit the movement of the endangered jaguarundi *(Herpailurus yaguarondi)* and ocelot on the refuge, hindering their ability to breed and to access water (some wall segments in Texas are being constructed as much as two miles north of the border and the Rio Grande). Fewer than 100 ocelots are believed to still survive in the United States, half of them within the Rio Grande valley.[9]

In addition, the border wall will have indirect environmental effects, by shifting flows of undocumented immigrants and subsequent border enforcement efforts to other portions of the border.[10] In this respect, impacts of the border wall continue a pattern of environmental degradation experienced since the early 1990s, when the U.S. Border Patrol began intensifying its security efforts in urban areas such as San Diego and El Paso under its "Southwest Strategy." Instead of stopping undocumented immigration, the Southwest Strategy has shifted migrant traffic to more remote areas of the border, in particular protected federal lands.[11] Indeed, levels of undocumented immigration into the United States steadily increased following the initiation of the Southwest Strategy.[12]

The shift in routes taken by undocumented immigrants into the United States has resulted in, apart from the increased toll on human lives, numerous adverse environmental effects—from both immigration and associated Border Patrol enforcement efforts—on wildlife and wildlife habitat, including habitat fragmentation, water pollution, soil damage and compaction, destruction of vegetation, and wildlife disturbance and displacement from light and noise (see chapters 6 and 14 in this volume).[13]

A Convergence of Responses

The initiation of border wall construction has engendered significant opposition, although legal options for challenge are greatly limited by the REAL ID Act waiver provision. In the San Pedro NCA case, conservation groups subsequently challenged the waiver provision as a violation of the Constitution's separation of powers principles. The district court judge, however, rejected these claims, and the U.S. Supreme Court refused to hear the case. Legal challenges to the method by which DHS is condemning private land for wall construction in Texas and a new constitutional challenge to the DHS secretary's recent waiver were also recently rejected.

The environmental threat posed by the wall has also brought renewed attention to the depth and diversity of transboundary conservation efforts. These efforts are illustrated by important binational treaties, agreements, and government programs such as the 1983 La Paz Agreement (U.S.–Mexico Agreement on Cooperation for the Protection and Improvement of the Environment in the Border Area), with its ensuing Integrated Border Environment Plan as well as the Border XXI and Border 2012 programs; the trilateral North American Free Trade Agreement and its associated Commission for Environmental Cooperation; Sister Parks, Wildlife Without Borders, and numerous other wildlife and invasive species–management programs; and the International Boundary and Water

Commission/Comisión Internacional de Límites y Aguas in its work to protect wetland systems along the Colorado River Delta. For example, under the U.S.–Mexico Sister Park Partnership, the U.S. National Park Service and Mexico's Comisión Nacional de Áreas Naturales Protegidas (National Commission of Protected Natural Areas) have designated seven sister parks in the borderland region (see chapter 13 in this volume).[14] The spirit and goals of these agreements, treaties, and collaborative programs are severely undermined by the unilateral wall construction being carried out by the United States.

Controversy surrounding the wall has, however, also helped revitalize the commitment of a growing number of scientists, academics, government officials, and advocates to increase collaboration with their cross-border partners. One example of this cooperation is the recent convergence between a group of U.S. and Mexican conservation leaders seeking to identify the potential impacts of wall construction and designate critical areas for cross-border wildlife movement. In the United States, this effort began in 2005 as a series of annual Border Ecological Workshops, organized by conservation organizations Defenders of Wildlife and the Wildlands Project, with the participation of U.S. stakeholders from state and federal wildlife agencies, DHS and Border Patrol, other conservation organizations, and academic representatives. The workshops resulted in two white papers and the goals of identifying critical cross-border wildlife corridors and "indicator" species and of developing recommendations for mitigating and protecting those corridors, particularly in Arizona.[15]

Concern about the potential environmental impacts of the border wall between the two countries was eventually expressed at the highest levels of Mexican government. The ministers of the Secretaría de Relaciones Exteriores (Secretariat of Foreign Relations) and the Secretaría de Medio Ambiente y Recursos Naturales (SEMARNAT, Secretariat of Environment and Natural Resources) identified the need to hold discussions with the United States on this topic and, in early 2007 the Mexican Senate officially requested that SEMARNAT analyze the situation. SEMARNAT organized an expert workshop in collaboration with El Colegio de la Frontera Norte, a respected research institution. The resulting Technical-Scientific Workshop on the Potential Environmental Impacts of the Border Fence Between the United States and Mexico—attended by more than fifty specialists from international agencies, civil society, academia, and state and federal government agencies on both sides of the border—was the first binational effort at tackling this problem and the first time a full border-length approach was explicitly taken. SEMARNAT, along with several partner organizations, published its proceedings in the book *A Barrier to Our Shared Environment: The Border Fence Between the United States and Mexico.*[16]

Both the Border Ecological Workshops in the United States and Mexico's workshops have evolved into cooperative, binational efforts to address the environmental dimensions of border wall construction and U.S. border security policy. This convergence will help ensure a continual binational effort to identify the potential impacts of border wall construction to lands and wildlife, with the goal of proposing defensible alternatives and mitigation strategies.

Policy Alternatives

As we have detailed in this chapter, the current border wall construction, being undertaken under broad waivers from environmental and other legal requirements, poses a significant threat to borderland protected lands and wildlife and strongly undermines transboundary conservation efforts. Border security and environmental protection are not mutually exclusive, however, and are in fact complementary in many respects. We present the following policy options to help better integrate environmental considerations into border security efforts.

Legislative Immigration Reform

Comprehensive immigration-reform proposals such as the U.S. Senate's 2006 bill S. 2611 attempt to establish a more orderly, lawful immigration system by creating a guest-worker program and providing for increased work and family visas. One of the main goals of such proposals is to reduce undocumented immigration, which would in turn decrease the need for environmentally damaging enforcement efforts.

In addition to this overall goal, S. 2611 contained several specific environmental provisions, including a requirement that DHS, in coordination with other agencies, assess the "international, national, and regional environmental impact" of wall construction. An amendment to the bill, sponsored by the late Senator Craig Thomas (R–Wyo.), required that the DHS secretary coordinate with the secretaries of interior and agriculture to develop a border-protection strategy that "best protects" federal lands, directed that all Border Patrol agents undergo natural-resource training coordinated with federal land-management agencies, and emphasized the use of low-impact technologies on federal lands. Finally, S. 2611 required extensive coordination with the Mexican government on a wide range of shared interests, including border security, human and drug trafficking, and assistance in the development of economic opportunities and job training for Mexican nationals. Many would take such cooperation a step farther, as evidenced by increasing calls for a "Mexican Marshall Plan."[17]

By attempting to address root causes of undocumented immigration, emphasizing cooperation with the Mexican government, and requiring that environmental considerations be integrated into border security efforts, broad immigration-reform efforts offer much greater opportunities for borderland environmental protection and the continued success of trans-boundary conservation initiatives than is provided for by current legislation and policy.

Interim Environmental Legislation

The ultimate success and timeline for passing comprehensive immigration reform remains highly uncertain given the deep polarization and strong opposition from some people in the United States. In recognition of the need to improve environmental protections now, Representative Raúl Grijalva (D–Ariz.) in June 2007 introduced the Borderlands Security and Conservation Act of 2007 (H.R. 2593). H.R. 2593 would incorporate the provisions of Senator Thomas's amendment described earlier; repeal the environmental waiver provisions of the REAL ID Act; create a borderlands conservation fund to provide financial assistance for projects intended to improve management of borderland species and habitat; mitigate the effects of border enforcement and undocumented immigration; and provide discretion to DHS to install "vehicle barriers" rather than impermeable fences in environmentally sensitive lands. Vehicle barriers, which are typically steel posts placed at intervals, are an important alternative to fencing because they are spaced so as to allow foot and animal traffic but not vehicular traffic. Representative Grijalva recently introduced a similar bill, H.R. 2076, in the 111th Congress.

Strengthen Alternative Avenues for Binational Cooperation

The scope and diversity of binational conservation efforts, which have strengthened in response to border wall proposals, demonstrate the interest that a wide diversity of stakeholders in the border region have in ensuring that environmental considerations are fully incorporated into border security measures. These efforts speak directly to Congress and DHS regarding the need to reevaluate current border security policy.

At the governmental level, in 2008 SEMARNAT and the State of California signed a Memorandum of Understanding on Environmental Cooperation addressing wildlife and habitat conservation, an agreement that might be extended to other border states. In addition, SEMARNAT and the Natural Resources Conservation Service of the U.S. Department of Agriculture are working on a memorandum with provisions for a binational borderlands restoration initiative.

At the nongovernmental level, conservation organizations on both sides of the California–Baja California border are joining in the Las Californias Binational Conservation Initiative, which seeks to preserve wildlife corridors along this section of the border and to convince lawmakers to reconsider a 32-kilometer (20-mile) stretch of mandated fence there.[18] Farther east, organizations such as Fundación Cuenca Los Ojos in Sonora are preserving land on the Mexican side of the border and implementing land-management strategies that are synergistic with those on the U.S. side of the border.

Private citizens and a significant segment of the private sector are also uniting binationally. The recently formed Binational Chamber of Commerce, including business leaders from southern Texas and Tamaulipas, has drafted resolutions against border fences, reflecting intense regional opposition to the fence.[19] In another example of action, the International League of Conservation Photographers has conducted a rapid assessment visual expedition, known as a RAVE, to make a baseline visual inventory of biodiversity, raise public awareness about the environmental implications of the border fence, and gain support for legislation changing the current fence policy.

Finally, as discussed earlier, researchers from government agencies, nongovernmental organizations, and academia have been actively identifying information needs and setting priorities for research on the potential environmental effects of the border wall in order to inform decision making.[20]

Conclusions

Congress has exempted U.S.–Mexico border wall construction from environmental and other federal laws, resulting in extensive damage to borderland wildlife and lands and the undermining of important and longstanding binational transboundary conservation efforts. There fortunately exist effective alternatives to this policy that can both further border security goals and preserve our shared natural environment.

Despite such alternatives, there are no easy solutions to the difficult intersection of undocumented immigration and antiterrorism concerns that are ostensibly addressed by current border security legislation and policy, or to their grave collateral environmental effects. In light of the continuing failure of U.S. border security to deter immigration effectively, however, it should be evident that a broader overhaul and reform of U.S. immigration policy will ultimately be needed to reduce flows of undocumented immigration and thus reduce the subsequent environmental damage created by such immigration and the efforts to deter it. Equally vital, the REAL ID Act's waiver provisions must be repealed, and the protections of U.S. environmental and cultural resource laws reinstated.

In addition, discussion, dialogue, and information gathering are essential and urgent elements of a solution strategy that encourages stronger dialogue and cooperation among parties. Congressional representatives from nonborder states need to visit, meet, and understand border stakeholders to have a better idea of what is happening on the ground. Dialogue must also be further strengthened among U.S. federal government agencies, between U.S. and Mexican government agencies, between newly elected U.S. president Barack Obama and Mexican president Felipe Calderón, between the U.S. and Mexican congresses, and with the public at large. Finally, this dialogue and discussion must be complemented by and based on site-specific scientific information regarding the borderlands environment and the effects of border security on that environment.

Our shared natural resources play a vital role in sustaining not only our rich heritage of wildlife, but also the economic health of communities all along the U.S.–Mexico border, a fact clearly reflected by the many cooperative binational efforts undertaken by many in those communities to preserve their irreplaceable and unique shared environment. U.S. border wall construction has threatened these efforts, but opposition to the wall continues to build, and existing transboundary conservation efforts will increasingly help illuminate the vital importance of conserving this unique and biologically rich area.

Notes

1. In this article, we use the terms *border wall* and *border fence* interchangeably. Although there are various types of border walls and fences being constructed along the southern border, all are approximately 12–15 feet (3.7–4.6 meters) in height and do not contain gaps wide enough for humans or wildlife to pass through. In contrast, "vehicle barriers" do contain gaps, allowing wildlife (and humans) to pass through.

2. T. Egan, "Republicans losing the West," *New York Times*, June 21, 2007.

3. J. Weisman, "Border fence is approved; Congress sets aside immigration overhaul in favor of 700-mile barrier," *Washington Post*, October 1, 2006.

4. The nonpartisan Congressional Research Service characterized the provision as "unprecedented," and the provision's unilateral grant of legislating authority to an executive branch official is widely viewed as a violation of the Constitution's fundamental separation of powers doctrine. See B. Nuñez-Neto and S. Viña, "Border security: Barriers along the U.S. international border," *Congressional Research Service*, December 12, 2006.

5. As one Congress member stated during the limited floor debate held on the bill, "A waiver this broad is unprecedented. It would waive all laws, including laws protecting civil rights; laws protecting the health and safety of workers; laws, such as the Davis-Bacon Act, which are intended to ensure that construction workers on federally funded projects are paid the prevailing wage; environmental laws; and laws respecting sacred burial grounds." Statement of Representative Sheila Jackson-Lee (D–Tex.), 109th Cong., 1st sess., February 9, 2005, 151 *Cong. Rec.* H 453, 459, daily ed.

6. U.S. Fish and Wildlife Service (FWS), *Biological Opinion: U.S.–Mexico Border Pedestrian Construction near Sasabe, Nogales, Naco, and Douglas, Arizona* (Washington, D.C.: U.S. FWS, 2007).

7. R. List, "The impacts of the border fence on wild mammals," and C. Varas, "Black bears blocked by the border," both in *A Barrier to Our Shared Environment: The Border Fence Between the United States and Mexico*, edited by A. Córdova and C. A. de la Parra, 77–86 and 87–91 (Mexico City and San Diego: Secretaría de Medio Ambiente y Recursos Naturales, Instituto Nacional de Ecología, El Colegio de la Frontera Norte, and Southwest Consortium for Environmental Research and Policy, 2007).

8. E. McCain and J. L. Childs, "Evidence of resident jaguars *(Panthera onca)* in the southwestern United States and the implications for conservation," *Journal of Mammalogy* 89(1) (2008): 1–10.

9. K. Menzer, "Border fence may stop ocelot," *Dallas Morning News*, June 24, 2007.

10. U.S. FWS, *Biological Opinion*.

11. J. Turner, *Transforming the Southern Border: Providing Security and Prosperity in the Post 9/11 World*, U.S. House Select Committee on Homeland Security report (Washington, D.C.: U.S. Government Printing Office, 2004).

12. J. Passel, *Unauthorized Migrants: Numbers and Characteristics* (Washington, D.C.: Pew Hispanic Center, 2005).

13. U.S. Government Accountability Office (GAO), *Border Security: Agencies Need to Better Coordinate Their Strategies and Operations on Federal Lands*, GAO Report 04-590 (Washington, D.C.: U.S. GAO, 2004); E. Marris, "Wildlife caught in crossfire of U.S. immigration battle," *Nature* 442 (2006): 338–39; B. Segee and J. Neeley, *On the Line: The Impacts of Immigration Policy on Wildlife and Habitat in the Arizona Borderlands*, report (Washington, D.C.: Defenders of Wildlife, 2006).

14. U.S. Department of the Interior, "Interior Deputy P. Lynn Scarlett and Undersecretary Felipe Adrian Vazquez of the Ministry for Environment and Natural Resources officially designate seven sister parks during 2006 U.S.–Mexico Bi-national Commission," press release, March 24, 2006, available at http://www .doi.gov/news/2006news.html, accessed August 31, 2007; U.S. National Park Service, "National Park Service, Office of International Affairs, frequently asked questions" (2007), available at http://www.nps.gov/oia/about/why.htm, accessed August 31, 2007.

15. The two white papers are entitled *Ecological Considerations for Border Security Operations* and *Alternatives and Mitigation for Border Security Infrastructure in Areas of Critical Ecological Concern.* Also see J. Vacariu and K. Neeley, "Alternatives and mitigation for border security infrastructure in areas of critical environmental concern: Outcomes and recommendations of the Border Ecological Workshop II," paper presented at the Border Ecological Workshop II, October 18, 2006, Tucson, Ariz.

16. Córdova and de la Parra, *A Barrier to Our Shared Environment.*

17. W. Cornelius, "Controlling 'unwanted' immigration: Lessons from the United States, 1993–2004," *Journal of Ethnic and Migration Studies* 31(4) (2005): 775–94.

18. Conservation Biology Institute, *Las Californias Binational Conservation Initiative: A Vision for Habitat Conservation in the Border Region of California and Baja California* (Corvallis, Ore.: Conservation Biology Institute, 2005); S. Dibble and M. Lee, "Tijuana sprawl cuts into key ecosystem: Border coalition hopes to save wildlife ecosystems," *San Diego Union-Tribune,* February 10, 2006.

19. S. Calderon, "Binational chamber working to regionalize border," *Brownsville Herald,* September 27, 2006.

20. Vacariu and Neeley, "Alternatives and mitigation for border security infrastructure"; Córdova and de la Parra, *A Barrier to Our Shared Environment;* and the two white papers: *Ecological Considerations for Border Security Operations* and *Alternatives and Mitigation for Border Security Infrastructure.*

Building
Institutional Bridges

Building Institutional Bridges

One of the greatest challenges of transboundary conservation is the fact that the administration of shared environments is divided between countries, each of which is composed internally of many jurisdictions, including municipalities, states, and government agencies at all levels. Further, the boundaries between countries and the cultural, political, and economic differences they entail may impede stakeholders' ability to work together across borders. Because of these administrative challenges, multilateral institutions are critical in building bridges across the divisions posed by the political line.

The two chapters in this section center on binational institutions tasked with coordinating the management of environmental resources shared by the United States and Mexico. Their mandates are often broader than conservation, but the institutions and arrangements do provide frameworks upon which conservation projects can be built. The manner in which the institutions are designed and the policies that support or hinder their conservation mandates are key components of their ability to foster successful transboundary conservation efforts. These chapters demonstrate the role that binational institutions currently play in conservation and point to strategies to strengthen this role. Although there are critical channels of collaboration at state and local levels, the chapters gathered in this section focus on formal, federal-level institutions.

In chapter 16, Steve Mumme and his coauthors discuss one such institution, the Commission for Environmental Cooperation (CEC). They argue that of all the multilateral arrangements at play across the U.S.–Mexico border, the CEC is the only institution with a mandate broad enough to encompass the continentwide spectrum of transboundary ecological and conservation concerns addressed in this volume. They analyze what they call the CEC's "SOS functions": *scaling* ecological linkages across North America, *organizing* networks of stakeholders, and *spotlighting* issues of transboundary environmental concern through investigations. Moreover, they demonstrate how biological conservation has become central to the CEC's core mission over its decade and a half of existence and highlight CEC's potential for incorporating a wide variety of stakeholders, including local people, into transboundary conservation. They also describe the obstacles faced by the CEC in fulfilling conservation functions, suggest ways that these obstacles might be overcome, and conclude by suggesting that the CEC be given more financial support.

In chapter 17, Ana Córdova and Carlos de la Parra examine the rich and diverse set of institutional arrangements that the U.S. and Mexican federal governments have developed to manage their shared borderlands environment. They trace a shift toward greater complexity and integration as well as the inclusion of a wider range of natural-resources issues in response to emerging environmental concerns and demands from both societies. They describe three institutional frameworks that have been used to coordinate between the two countries, and they test each framework's resilience and effectiveness against the issue of the U.S. construction of border security infrastructure along the international boundary. They conclude that up to January 2009, these frameworks were underutilized. However, they end on a hopeful note—suggesting that the ideals of international cooperation expressed by the Barack Obama administration may signal an era of renewed U.S.–Mexico cooperation to manage their shared environment.

The Commission for Environmental Cooperation and Transboundary Conservation across the U.S.–Mexico Border

Stephen P. Mumme, Donna Lybecker, Osiris Gaona, and Carlos Manterola

In a Nutshell

- The Commission for Environmental Cooperation (CEC) was created by the environmental side agreement to the North American Free Trade Agreement.
- The CEC's trinational mandate includes regional (U.S., Mexican, and Canadian) environmental protection, trade protection, sustainable development, and species conservation.
- The CEC's core functions are *scaling* ecological linkages across North America, *organizing* networks of conservation participants, and *spotlighting* matters of environmental concern through investigations.
- Four case studies demonstrate the CEC's capacity for transboundary conservation efforts: the monarch butterfly, the North American bat, the Sonoran pronghorn antelope, and the San Pedro River riparian corridor.
- However, the CEC's limited funding from member nations does not permit it to realize its mandate fully.
- The CEC is uniquely situated to meet transnational conservation requirements, remains an important tool for regional efforts, and should be given more support in the future.

Introduction

Biodiversity conservation across the U.S.–Mexico border presents an administrative challenge that exceeds the capacity of either national government. The North American Commission for Environmental Cooperation (CEC), established by a side agreement to the North American Free Trade Agreement, is an important institutional player supporting transboundary conservation initiatives. Although the CEC has struggled with limited support from the three governments involved, it occupies a unique and valuable niche in North American transboundary conservation that is particularly vital for conservation across the U.S.–Mexico border and deserves greater and sustained trinational support.

Of all the arrangements affecting transboundary conservation across the U.S.–Mexico border, the North American Agreement on Environmental Cooperation (NAAEC) has the most potential for strengthening international cooperation across the wide spectrum of ecological concerns. It certainly has the most expansive international mandate in this issue area. However, since the NAAEC entered into force on January 1, 1994, its institutional arm, the CEC, has struggled for survival, often neglected by its member governments. Within this adverse administrative environment, the CEC has managed to carve out an institutional role in transboundary conservation policy at the regional level that is vital for supporting conservation initiatives along and across the U.S.–Mexico border.

The CEC and Transboundary Conservation

The CEC is the only international institution with a truly regional and comprehensive mandate to protect the North American environment. Conservation is technically only one of the CEC's environmental concerns, yet after more than a decade and a half of its operation, biological conservation has become central to its vision and role in advancing environmental cooperation in North America. The CEC's organizational logo, the monarch butterfly, symbolizes its commitment to conservation of the natural resources on which the biodiversity of North America depends. Even as the CEC has pared its operational priorities in the face of acute institutional and financial limitations—as discussed more fully later—conservation remains its core mission.

The CEC's role in regional and transboundary conservation revolves around what may be described as its SOS functions: *scaling, organizing,* and *spotlighting.* As the only environmental agency with a trinational mandate, the CEC is able to focus on ecological issues that matter at a North American scale—addressing ecological concerns that link the entire region. With its unique mandate for advancing regionwide cooperation, it has an important role in convening governmental, scholarly, and advocacy networks for science-based policy action on transboundary conservation. Through its investigative authority, it has the special capacity to study and publicize critical biodiversity issues of transboundary and regional concern. As former executive director Greg Block points out, the CEC can shine a spotlight on matters of critical environmental concern for the trinational region.[1]

The CEC and Conservation across the U.S.-Mexico Border

The value of these SOS functions is particularly evident at the U.S.–Mexico border. The boundary runs 3,141 kilometers (1,952 miles) from coast to coast bisecting distinctive ecological zones and transecting the species migration corridors that link the Arctic with Central and South America. It cuts through two of North America's biodiversity "hot spots"—the California Floristic Province and the Madrean Pine–Oak Woodlands of the Sonoran and Chihuahuan desert zones. The California Floristic Province alone contains an estimated 3,488 species of plants, 61 percent of which are native to the region.[2] The boundary either abuts or passes near thirty-seven federal protected areas in the United States and seven in Mexico, which represent billions of dollars of conservation investment between the two nations.[3]

Scaling

Although the burden of managing this extraordinary abundance of fauna and flora falls almost entirely on the Canadian, U.S., and Mexican governments' domestic agencies, the CEC directs attention to the scale and significance of these ecological linkages across North America. It does so through its annual *State of the North American Environment Report*, detailing the ecological regions of North America; strategic planning for biodiversity protection—for example, action plans for critical marine regions, including the Bight of California and the Gulf of Maine; its North American bird-conservation initiative; and the targeting of species of common conservation concern for priority attention.[4]

In December 2002, the CEC released its Strategic Plan for North American Cooperation in the Conservation of Biodiversity, which is aimed at promoting conservation of regions of continental ecological significance; promoting conservation of migratory and transboundary species; facilitating data and information sharing as well as integrated biodiversity monitoring; facilitating networking, sharing of best practices, education, and training; promoting collaborative responses to common threats to North American ecosystems, habitats, and species; and identifying collaborative opportunities for biodiversity conservation and sustainable use of resources arising from free trade.[5] Under this mantle of programs, the CEC prioritized seventeen

Species of Common Conservation Concern, all of which are found near or migrate across the U.S.–Mexico border (see table 16.1).[6] Three species— the Western Burrowing Owl *(Athene cunicularia)*, the Ferruginous Hawk *(Buteo regalis)*, and the black-tailed prairie dog *(Cynomys ludovicianus)*— have since been targeted with conservation action plans.

Organizing

The CEC has functioned as a catalyst and venue for developing and organizing knowledge-based networks of conservation scientists, policy specialists, and conservation advocates to focus on transboundary conservation issues in North America. As noted, it aims to facilitate and promote the trinational exchange of information and research, with particular emphasis on areas of common conservation concern. It has organized and supported a broad range of networks, established working conservation partnerships, and coordinated efforts to develop regional conservation strategies.[7] Its North American Biodiversity Information Network, for example, is supporting rigorous scientific understanding of conservation issues and making this information widely available to scholarly, governmental, and advocacy communities.[8] The CEC's recently released North American Monarch Conservation Plan, for instance, emerged from a 2007 CEC-supported workshop in Morelia, Michoacán, and the CEC's Trilateral Monarch Butterfly Sister Area Network.[9] The black-tailed prairie dog action plan emerged from a 2004 CEC workshop in Calgary and a CEC-organized network of experts.[10] Similar networks specific to the U.S.–Mexico border have been established for other species, including the trilateral committee for the lesser and Mexican long-nosed bats (see case studies presented later).

Spotlighting

The CEC's authority to investigate compelling matters of environmental concern for the North American region, found in NAAEC Articles 13 and 14, enables it to spotlight critical problems and shape government policy agendas. Its Article 13 authority, allowing it to initiate investigations independently, has been particularly important for transboundary conservation and the U.S.–Mexico border region. The commission has used this authority sparingly, but with real effect. Two of its six investigations have centered on transboundary conservation. Its first Article 13 investigation examined migratory waterfowl deaths at Silva Reservoir in the Mexican state of Guanajuato and recommended resource-management reforms that the governments have largely embraced.[11] The report based on its second

Table 16.1. North American Terrestrial Species of Common Conservation Concern.

black bear	*Ursus americanus*
black-tailed prairie dog	*Cynomys ludovicianus*
Burrowing Owl	*Athene cunicularia*
California Condor	*Gymnogyps californianus*
Ferruginous Hawk	*Buteo regalis*
Golden-cheeked Warbler	*Dendroica chrysoparia*
gray wolf	*Canis lupus*
(greater) Mexican long-nosed bat	*Leptonycteris nivalis*
lesser long-nosed bat	*Leptonycteris curasoae yerbabuenae*
Loggerhead Shrike	*Lanius ludovicianus*
Mexican Spotted Owl	*Strix occidentalis lucida*
Mountain Plover	*Charadrius montanus*
Northern Spotted Owl	*Strix occidentalis caurina*
Peregrine Falcon	*Falco peregrinus*
Piping Plover	*Charadrius melodus*
Sonoran pronghorn	*Antilocapra americana sonoriensis*
Whooping Crane	*Grus americana*

Source: Commission for Environmental Cooperation, "Biodiversity Conservation, Conservation of Migratory and Transboundary Species," working draft, Montreal, Canada, October 18, 2000.

investigation, *Ribbon of Life*, details the threatened ecology of the San Pedro River that crosses the Arizona-Sonora border.[12] The study, which many local interests initially resisted, led to the formation of water-management and wildlife partnerships for river conservation among a wide range of binational stakeholders.[13]

The CEC's regulatory capacity also springs from its Article 14 obligation to act on credible citizen-initiated allegations of governmental failure to enforce national environmental laws, including treaty obligations.[14] Few of the more than fifty citizen submissions have focused on the U.S.–Mexico border, however, and only one of fifteen factual records resulting from these complaints, involving contamination of the Río Magdalena in northern Sonora, centers on border area natural-resource management.[15]

The opportunity for citizens to monitor and hold governments accountable for the enforcement of conservation regulations is nevertheless a useful tool in the arsenal of transboundary conservation measures along the border.

The CEC is well positioned for advancing transboundary conservation measures for the border region and North America. Its SOS functions are vital, framing issues on a North American scale, organizing and promoting rigorous scientific investigation and broader public awareness of North America's transboundary ecosystems, and directing public attention to neglected issues of transnational interest. Its ability to promote cross-regional study and policy awareness of migratory species across boundaries is perhaps its most important conservation function. The CEC has the capacity to identify critical transboundary wildlife issues, support existing intergovernmental and scientific partnerships, and elevate the issues on national and trinational policy agendas by supporting and publicizing these initiatives.

Four Case Studies

The CEC's conservation impact along the U.S.–Mexican border affects numerous species of fauna and flora throughout the region, including the seventeen prioritized Terrestrial Species of Common Conservation Concern. A closer look at several of these cases illustrates the importance of the commission's SOS functions and its special capacity for strengthening bilateral conservation initiatives.

The Monarch Butterfly

The monarch butterfly *(Danaus plexippus)* may well be the most recognizable and charismatic insect species in North America, known for its rich, black-veined orange wings and its remarkable multigeneration annual migration from the highlands of central Mexico to the southern reaches of Canada. The milkweed-feeding monarch's extraordinary vulnerability across its epic trek led the International Union for the Conservation of Nature and Natural Resources (IUCN) in 1983 to designate its winter roosts in Mexico and California a threatened phenomenon, the first such designation in the IUCN's history.

Although domestic wildlife agencies and conservationists in Canada, the United States, and Mexico have actively sought to protect the species for more than thirty years, the IUCN designation drew attention to the need for better international coordination. At its inception, the CEC was viewed as a logical umbrella for promoting monarch conservation on the regional level.

In 1996, at the initiative of national wildlife agencies, the Trilateral Committee for Wildlife and Ecosystem Conservation and Management (hereafter, Trilateral Committee for Wildlife) was established by the three governments to support projects and programs and to build working conservation partnerships on issues of transboundary concern.[16] In 1997, the CEC and the U.S. Fish and Wildlife Service followed up by convening stakeholders in Morelia, Michoacán, to develop a long-range monarch-conservation strategy. The CEC's Biodiversity Conservation Working Group developed further bilateral cooperation and discussions between U.S. and Mexican agencies.

Although the Trilateral Committee for Wildlife and domestic agencies have prioritized monarch conservation, the need for an international body to help coordinate the many efforts across the North American region was evident. When in 2006 U.S. state and federal agencies mounted a new initiative to develop a cohesive protection plan, the CEC supported the development of the North American Monarch Conservation Plan. The CEC, working with the Trilateral Committee for Wildlife and other agencies, convened another meeting of experts in Morelia, Michoacán, in 2007. Its product was the first definitive North American Monarch Conservation Plan, published in 2008, coupled with other collaborative initiatives to monitor and protect threatened monarch habitat throughout North America.[17]

Several elements of the new plan are centered on the border region. As part of the monarch's flyway, the border region figures in the plan's flyway threat-mitigation strategy. Specific actions include identifying habitats essential for roosting and nectaring sites as well as developing guidelines for habitat conservation, enhancement, and restoration. The border region also figures in the plan's innovative enabling approaches in several ways, most specifically in a proposed action to expand the Sister Protected Area Network from the Gulf of Mexico to as far west as Amistad National Recreation Area on the Rio Grande.[18]

North American Bats

Bat species are among the most common mammalian species both worldwide and in North America. Despite the fact that they provide vital ecosystem services as pollinators, seed dispersers, and insect predators, bats have long been neglected by official conservation agencies. Today, North America's bat populations are under severe stress from human activity, growth, and climate change. As many as ten species of bats are known to cross the U.S.–Mexico border, contributing valuable services to ecosystems and agriculture (see chapters 9 and 11 in this volume).

Although bat conservation has been on the radar screen of U.S. and Mexican ecologists for some time, it has failed to gain traction at the government level. The Trilateral Committee for Wildlife, supported by the CEC, was instrumental in creating the North American Bat Conservation Partnership in 1998, but this initiative failed to develop.[19]

Government wildlife agencies play the leading role in the Trilateral Committee for Wildlife and control its resources, but the CEC has proved a valuable mechanism for keeping bat conservation before the public and highlighting its importance for the North American region. In 2002, as part of its Terrestrial Species of Common Conservation Concern initiative, it listed two species of transboundary bats, the Mexican long-nosed and lesser long-nosed species *(Leptonycteris curasoe* and *L. nivalis)*, as North American conservation priorities.[20] In 2005–2006, concerned ecologists raised the need to revive the North American Bat Conservation Partnership within the Trilateral Committee for Wildlife, lobbying for a new North American bat conservation initiative to monitor bat populations at a continental scale and coordinate conservation strategies.

The CEC has the capacity to endorse, convene, and facilitate such bilateral and trilateral partnerships among government agencies and scholars crafted around the pursuit of sound scientific knowledge of conservation challenges. As scientific consensus emerges from the bat initiative and other such partnerships, the CEC's ability to spotlight and publicize these priorities at a North American scale is an essential component of mobilizing political support and requisite funding for the protection of bats and other at-risk transboundary species.

The Sonoran Pronghorn Antelope

The Sonoran pronghorn antelope *(Antilocapra americana sonoriensis)*, a listed endangered species in the United States since 1967 and an endangered species in Mexico since 2000, inhabits a narrow range bridging the western boundary of Arizona and the Mexican state of Sonora. In Arizona, the pronghorn is found mainly in Organ Pipe Cactus National Monument and other federal wildlife and military reserves abutting the border; in Sonora, it ranges west of the small municipality of Sonoyta and near the Sierra de Pinacate, and as far south as the resort and fishing village of Puerto Peñasco.

Habitat for the Sonoran pronghorn has been severely restricted over the past half century, its movements hemmed in by myriad fences, highways, and barriers. The chain-link fence along Mexico's Highway 2 tracing the international boundary was identified more than a decade ago as a significant barrier to north–south movement and dispersal, a problem that has worsened with intensive road and barrier construction associated with the recently authorized U.S. boundary fence (see chapter 15 in this volume).

Conservation of the Sonoran pronghorn is linked to the conservation of antelope throughout North America. Although endangered in Mexico and along the U.S.–Mexico border, the pronghorn antelope *(Antilocapra americana)* is a species with healthy populations in Canada and the United States, where it is considered a game species. This contrasting situation is also an opportunity for collaborative work to recover the Sonoran pronghorn in Mexico and along the U.S.–Mexican border.

Federal and state wildlife managers have long been concerned with pronghorn preservation, although private conservation groups, or nongovernmental organizations (NGOs), have often taken the lead in its protection. In 1990, the Campfire Conservation Fund created the International Sonoran Antelope Foundation to generate support for recovery efforts. Defenders of Wildlife, the International Sonoran Desert Alliance, the Sonoran Institute, the Arizona-Sonora Desert Museum, and other groups are also active in its conservation. Since the mid-1990s, the Mexican group Unidos Para la Conservación and the New Mexico Department of Game and Fish, with the support of Mexican and U.S. federal authorities, have cooperated to reintroduce pronghorns to the state of Coahuila.

These initiatives are largely the result of the work of the Trilateral Committee for Wildlife, which identified pronghorn antelope conservation as a North American conservation priority in its first meeting in April 1996. In 1997, the Mexican and U.S. governments agreed to establish joint study and protection programs aimed at pronghorn antelope conservation.[21] Under these auspices, federal and state governments in the two countries have joined forces for pronghorn protection. One such initiative supported by New Mexico state authorities, for example, aims at reintroducing a group to the Mexican state of San Luis Potosí; another group has been reintroduced on Tiburon Island in the Sea of Cortés (Gulf of California). Along the Arizona-Sonora border, the two countries are also cooperating to restore the Sonoran pronghorn. In this area, a captive-breeding program at Cabeza Prieta National Wildlife Refuge has seen the local population of pronghorns reach nearly one hundred animals from a low of twenty-one in 2002.[22]

The CEC has been valuable in elevating awareness of the pronghorn's plight. Although the two governments and NGOs have taken the lead in pronghorn protection, the Trilateral Committee for Wildlife, working with the CEC, placed a spotlight on the species by listing it as a Terrestrial Species of Common Conservation Concern in 2003. The species was not selected as one of the three species warranting a North American Conservation Action Plan in 2004, but the CEC continued to support the efforts of the committee's biodiversity working group in monitoring the pronghorn.

In 2007, the U.S. decision to build security fencing along the boundary spurred the Trilateral Committee for Wildlife to focus on the problem of connectivity and physical barriers at its meeting in Quebec City. At the 2008 trilateral meeting in Veracruz, Mexico, the committee determined to redouble its pronghorn commitment, advancing a three-year work plan. The new plan sustains a North American commitment to Sonoran pronghorn recovery, monitoring of the peninsular pronghorn, and an ambitious project of further reintroductions of pronghorns in the Mexican interior.

In sum, the institutional partnership between the Trilateral Committee for Wildlife and the CEC has proven vital for advancing governmental and NGO initiatives and partnerships for protecting this important species in and across the border area. The trilateral meetings are instrumental in this collaborative process for the conservation and recovery of this shared species, one of the most charismatic species found on and across the U.S.–Mexico border. The CEC plays a vital role in legitimizing and promoting this objective.

The San Pedro River Riparian Corridor

The San Pedro River, with headwaters south of the Sonoran mining town of Cananea, runs for nearly 224 kilometers (140 miles) to the northwest, crossing the international boundary 40 kilometers (25 miles) north of Cananea, just west of the twin border villages of Naco, Arizona, and Naco, Sonora. From there, it bisects Arizona's Huachuca and Mule Mountains near Sierra Vista, Arizona, and continues to its confluence with the Gila River near Winkleman, Arizona. Much of the stream is ephemeral, but for a short 28.9-kilometer (18-mile) stretch between the Arizona sites of Hereford and Fairbank the flow is perennial, one of the few such water sources in the Sonoran Desert. This free-flowing river sustains the largest riparian willow-cottonwood groves in the Southwest.

The San Pedro River corridor supports an extraordinary diversity of fauna and flora, rivaling Costa Rica in its diversity of animal species. More than 350 species of migratory birds use its forested banks as they transit from breeding areas in Canada and the United States to wintering habitats in Mexico, Central America, and South America.[23] In 1988, the U.S. Congress designated a 65-kilometer (40-mile) portion of the San Pedro riparian system as the San Pedro River National Conservation Area, charging the Bureau of Land Management with its stewardship.

Despite its importance, the river's trove of ecological values has been at risk. Numerous ranches, farms, and human settlements on both sides of the border draw on its basin for water, including the rapidly growing city of Sierra Vista, Arizona, and a large U.S. military base nearby, Fort Huachuca. Environmentalists and government land managers have often clashed with local governments and private stakeholders over how to manage the river basin sustainably.

In 1997, recognizing the area's critical role as a habitat for migratory birds in North America, the CEC commissioned an expert study of the binational riparian area, the Upper San Pedro Initiative. The objectives were to initiate a process for stakeholder involvement in the development of sustainable strategies for preserving the ecosystem, to create a model of cooperation for other transboundary basins, and to inform the public about the regional importance of preserving migratory bird habitat. The CEC's Upper San Pedro Advisory Panel report helped energize local conservation initiatives on the U.S. side and catalyzed binational awareness of the issue.[24]

Inspired in good measure by the CEC's report, the Upper San Pedro Partnership was formed in 1998 to build consensus and support for sustainable policies and practices among stakeholders in the Sierra Vista Subwatershed. This collaboration of more than twenty participating organizations—including The Nature Conservancy, U.S. Army Fort Huachuca, the City of Sierra Vista, and the U.S. Geological Survey—aims to ensure that an adequate long-term groundwater supply is available to meet the current and future needs of area residents as well as the San Pedro River National Conservation Area.[25] The partnership has succeeded in advancing urban water conservation and preserving open space on the area perimeter, including development of an innovative conservation easement and the adoption of new land-use practices at the military base. The partnership has also spearheaded efforts to work with stakeholders in the Mexican reach of the watershed.

These four cases illustrate well the importance of the CEC's SOS functions in addressing transboundary conservation challenges in the U.S.–Mexican border region. In each case, the CEC's scaling function situates these issues and justifies their importance in a North American frame. The CEC has taken the initiative and used its capacity to organize responses at a trilateral level, even when the problem, as in the case of the Sonoran pronghorn, is largely bilateral in nature. In addition, it has drawn on its mandate and

Table 16.2. Objectives of the North American Agreement on Environmental Cooperation.

- Protect North American environment for present and future generations
- Promote sustainable development
- Increase cooperation to conserve and protect the environment, including flora and fauna
- Support NAFTA's environmental goals
- Avoid creating trade distortions or barriers
- Strengthen cooperation on the development and improvement of environmental laws, procedures, and practices
- Enhance compliance and enforcement of environmental laws and regulations
- Promote transparency in development of environmental laws and policy
- Promote economically efficient and effective environmental measures
- Promote pollution-prevention policies and practices

Source: "North American Agreement on Environmental Cooperation," signed September 13, 1993, Art. 1, available at http://www.cec.org/pubs_info_resources/law_treat_agree/naaec/index.cfm?varlan=english.

resources to spotlight these issues at the regional and international level, thus legitimizing and strengthening public and government concern for these problems. Unfortunately, though, the CEC has itself struggled for survival, putting its conservation mandate and its efforts to protect transboundary wildlife at risk.

The CEC Quandary: Mandate, Organization, and Trajectory

From its inception, the CEC has struggled with an extraordinary mandate that is both a blessing and a curse. Embracing the core principles of sustainable development and environmental protection expressed in the Brundtland Report and the United Nation's Agenda 21, the NAAEC commits the CEC to a comprehensive agenda that spans regional environmental protection, trade protection, sustainable development, species conservation, improvement of environmental laws, strengthened cooperation on enforcement and regulatory compliance, promotion of transparency and public participation, and promotion of pollution prevention (see table 16.2).[26]

Table 16.3. Commission for Environmental Cooperation: areas of concern.

- Data gathering, analysis, and sharing
- Border pollution prevention
- State of the environment indicators
- Economic instruments for environmental objectives
- Environmental research and technology development
- Public environmental awareness
- Exotic species
- Conservation of flora and fauna habitat and of protected areas
- Protection of endangered and threatened species
- Environmental emergency preparedness and response
- Environment and economic development
- Human resource in environmental fields
- Scientist exchange
- Environmental compliance and enforcement
- Ecologically sensitive national accounts
- Ecolabeling
- Other matters as it may decide

Source: "North American Agreement on Environmental Cooperation," signed September 13, 1993, Art. 1, available at http://www.cec.org/pubs_info_resources/law_treat_agree/naaec/index.cfm?varlan=english.

Guided by these objectives, the NAAEC's Article 10(2) authorizes the CEC's governing council to consider and develop recommendations across a broad range of environmental concerns (table 16.3). Conservation and transboundary issues are an important part of this agenda. In addition, the NAAEC tasks the CEC to monitor the enforcement of domestic environmental law in the North American region and make provisions for both citizen-initiated and secretariat-initiated investigations of alleged governmental failure to enforce environmental legislation in North America. These regulatory functions remain controversial.

The NAAEC established a tripartite institutional structure to administer this mandate: the Council of Ministers (the environmental ministers of Canada, the United States, and Mexico), the CEC secretariat, and the Joint Public Advisory Council consisting of business, government, and environmental representatives from the three countries. From the outset, the CEC secretariat found itself wedged in by high public expectations, a nearly impossible set of mandate objectives, and limited support from its member governments—best reflected in a $9 million budget that has remained unchanged since 1994. The result has been a protracted search for a reliable niche in North American environmental management.[27]

The CEC's quest for a reliable policy niche supported by its member governments is one of the important factors leading to a greater emphasis on conservation and wildlife protection. After an initial period of experimentation with its many priorities, the CEC was forced to scale back its work program, downsizing radically around the twin themes of promoting "environmental sustainability in open markets" and "stewardship of the North American environment."[28] The CEC's work program has since centered on four program areas: environment and trade; conservation of biodiversity; pollutants and health; and law and policy.

After its ten-year review in 2004, the CEC saw further program shrinkage. Its new long-term vision, set out in its June 2004 Puebla Declaration, views the CEC as a catalyst for change in partnership with other governmental entities and stakeholders within the North American region.[29] According to this new vision, the CEC would facilitate regional action, generate concrete policy results based on rigorous analysis and policy recommendations in specific program areas, and provide high-quality compatible data at a North American scale to support sound environmental science.[30]

To implement this new, leaner, and more efficient vision of the CEC's role, a further paring of its priorities and programs was undertaken in 2005, folding activities in the cooperative work program into three new overarching priorities: (1) information for decision making, (2) capacity building, and (3) trade and environment.[31] This streamlining shrank CEC's programs to fifteen such project initiatives (see table 16.4). Conservation broadly construed is incorporated directly into four of these project areas, including mapping North American environmental issues, reporting on the state of the North American environment, strengthening wildlife enforcement capacity, and building local capacity for integrated ecosystem management and the conservation of critical species and spaces.

Table 16.4. Commission for Environmental Cooperation cooperative work programs, 2007–2009.

Priority 1: Information for Decision-Making Projects

- Monitoring and assessing pollutants across North America
- Tracking pollutant releases and transfers in North America
- Enhancing North American air-quality management
- Mapping North American environmental issues
- Reporting on the state of the North American environment

Priority 2: Capacity-Building Projects

- Strengthening wildlife-enforcement capacity
- Improving private- and public-sector environmental performance
- Building local capacity for integrated ecosystem management and to conserve critical species and spaces
- Sound management of chemicals

Priority 3: Trade and Environment Projects

- Promoting the North American renewable energy market
- Encouraging green purchasing
- Harnessing market forces for sustainability
- Encouraging trade and enforcing environmental laws
- Developing guidelines for risk assessment of invasive alien species and their pathways
- Undertaking ongoing environmental assessment of the North American Free Trade Agreement

---

Source: Commission for Environmental Cooperation (CEC), *Operational Plan of the Commission for Environmental Cooperation,* 2007–2009 (Montreal: CEC, February 6, 2007), 3.

Conclusions

Since its founding, the CEC has struggled to confirm its place in North American environmental management. Advancing trinational cooperation for conservation of North America's ecosystems nevertheless has remained a core element of its mission even as its effective mandate has narrowed in response to institutional and fiscal constraints. Its SOS functions are critical for generating transnational action on transboundary conservation issues of common concern to the North American community.

Along the U.S.–Mexican border, the CEC has generated needed scientific study and public awareness of the important role and many ecological services that border-area ecosystems provide across the gamut of North American biodiversity. The CEC's importance is evident in each of the four cases reviewed earlier. Its SOS functions effectively situate the border within a regionwide ecology; stimulate organizational responses in the form of expert research, stakeholder advocacy, and governmental support for the sustainable management of terrestrial fauna and flora; and galvanize public response by investigating conservation problems of concern to the North American community.

The importance of these functions is very much apparent at a time of heightened adverse impacts of human activities at or near the U.S.–Mexico border. Growth-related depletion of water resources is stressing border ecosystems as never before, threatening wildlife habitat from the Colorado River delta to the Rio Grande. Invasive species compete with native vegetation for riparian resources. Heightened national-security and immigration infrastructure along the U.S.–Mexican border severely threaten border ecosystems with impacts that have not yet been properly studied or adequately understood.[32]

Although the CEC continues to be limited by government will, its mandate and role in North American environmental management and wildlife conservation have never been more critical. It is uniquely suited to take stock of human impacts on our shared biodiversity and to direct attention to the need for stewardship of natural resources in North America. No other national or binational agency has its SOS capacity. Along the U.S.–Mexican border, the CEC provides some degree of insurance that critical biodiversity issues are less apt to be marginalized and disconnected from the North American main. Its ability to do so, as John Knox and David Markell aptly put it, "depends directly on the actions of individuals, groups, and governments willing to pursue its goals."[33] The governments are especially critical to this mix. If its work in transboundary conservation to date is any measure, the CEC deserves a greater measure of their support going forward.

Notes

1. G. Block, "The CEC Cooperative Program of Work," in *Greening NAFTA*, edited by D. Markell and J. Knox, 25–37 (Palo Alto, Calif.: Stanford University Press, 2003).

2. H. Riemann, "Ecological risks involved in the construction of the border fence," in *A Barrier to Our Shared Environment: The Border Fence Between the*

United States and Mexico, edited by A. Córdova and C. de la Parra, (Mexico City and San Diego: Secretaría de Medio Ambiente y Recursos Naturales, Instituto Nacional de Ecología, El Colegio de la Frontera Norte, and Southwest Consortium for Environmental Research and Policy, 2007), 105.

3. U.S. Department of Interior, *Fact Sheet: Natural and Cultural Resource Areas along the U.S.–Mexico Border* (Washington, D.C.: U.S. Department of Interior, n.d.), available at http://137.227.231.90/FCC/resource-areas.htm, accessed April 30, 2008.

4. Commission for Environmental Cooperation (CEC), *Conservation of Biodiversity* (Montreal: CEC, 2002), available at http://www.cec.org/files/PDF/BIODIVERSITY/conserv_bio03-05_en.pdf, accessed July 10, 2008.

5. CEC, *Strategic and Cooperative Action for the Conservation of Biodiversity in North America* (Montreal: CEC, 2002), available at http://www.cec.org/files/pdf/BIODIVERSITY/211-03-05_en.pdf, accessed July 1, 2008.

6. CEC, "Biodiversity Conservation, Conservation of Migratory and Transboundary Species," working draft, Montreal, October 18, 2000.

7. Block, "The CEC Cooperative Program of Work," 35.

8. CEC, *North American Biodiversity Information Network* (Montreal: CEC, 2002), available at http://www.cec.org/programs_projects/conserv_biodiv/ project/index.cfm?projectID=21&varlan=english, accessed July 1, 2008.

9. CEC, *North American Monarch Conservation Plan* (Montreal: CEC, 2008), available at http://www.cec.org/pubs_docs/documents/index.cfm?varlan= english&ID=2300, accessed June 27, 2008.

10. CEC, *North American Conservation Action Plan: Black-tailed Prairie Dog* (Montreal: CEC, 2005), available at http://www.cec.org/files/pdf/BIODIVERSITY/NACAP-BlackTailed-PraireDog_en.pdf, accessed July 2, 2008.

11. CEC, *Silva Reservoir* (Montreal: CEC, 1995), available at http://www.cec.org/files/pdf//silv-e_EN.pdf, accessed January 18, 2003.

12. CEC, *Ribbon of Life* (Montreal: CEC, 1999), available at http://www.cec.org/files/PDF//sp-engl_EN.pdf, accessed July 12, 2008.

13. D. Tarlock and J. Thorson, "Coordinating land and water use in the San Pedro River basin," in Markell and Knox, eds., *Greening NAFTA*, 217–36.

14. R. Glicksman, "The CEC's biodiversity conservation agenda," and D. Markell, "The CEC citizen submissions process: On or off course?" both in Markell and Knox, eds., *Greening NAFTA*, 57–79 and 274–98.

15. S. Mumme and D. Lybecker, "The Commission for Environmental Cooperation in North America: Lessons for export to the hemisphere?" paper presented at the International Congress of the Latin American Studies Association, Montreal, September 10, 2007.

16. "Memorandum of Understanding to Establish the Canada/Mexico/United States Trilateral Committee for Wildlife, Plants, and Ecosystem Conservation and Management" (replaces the Joint Committee of 1995 and the Tripatriate Committee of 1988), Oaxaca, Mexico, 1996, available at http://www.trilat.org/general_pages/tri_mou.pdf, accessed July 15, 2008.

17. CEC, *North American Monarch Conservation Plan* (2008).

18. Ibid., 40.

19. "Memorandum of Understanding to Establish the Canada/Mexico/United States Trilateral Committee for Wildlife"; CEC, *North American Bird Conservation Initiative* (Montreal: CEC, 2002), available at http://www.cec.org/files/pdf/ BIODIVERSITY/221-03-05_en.pdf, accessed July 14, 2008.

20. CEC, "Biodiversity Conservation, Conservation of Migratory and Transboundary Species."

21. Ibid., 44.

22. A. Rotstein, "Pronghorn numbers show sign of bouncing back," *Arizona Daily Star*, December 2, 2008.

23. The Nature Conservancy, *San Pedro River, Arizona* (Arlington, Va.: The Nature Conservancy, 2008), available at http://www.nature.org/initiatives/fresh water/work/sanpedroriver.html, accessed August 12, 2008.

24. CEC, "North American Agreement on Environmental Cooperation (NAAEC)," *North American Environmental Law and Policy* (Winter 1998): 3–48.

25. A. Browning-Aiken, H. Richter, D. Goodrich, B. Strain, and R. Varady, "Upper San Pedro Basin: Fostering collaborative binational watershed management," *Water Resources Development* 20(3) (September 2004): 353–67.

26. G. Bruntland, ed., *Our Common Future: The World Commission on Environment and Development* (Oxford, U.K.: Oxford University Press, 1987); United Nations Sustainable Development (UNSD) and United Nations Conference on Environment and Development (UNCED), *Agenda 21* (Rio de Janeiro: UNSD and UNCED, 1992).

27. G. Hufbauer and J. Schott, *NAFTA Revisited: Achievements and Challenges* (Washington, D.C.: Institute for International Economics, 2004).

28. CEC, "CEC Council Joint Communiqué," Fifth Regular Session of Council, Merida, Mexico, June 26, 1998, at 4–7.

29. CEC, "Puebla Declaration," Eleventh Regular Session of the CEC Council, Puebla, Mexico, June 23, 2004.

30. CEC, *North American Conservation Action Plan: Black-tailed Prairie Dog;* CEC, *Ten Years of North American Environmental Cooperation* (Montreal: CEC, June 15, 2004).

31. CEC, *North American Conservation Action Plan: Black-tailed Prairie Dog,* 9.

32. C. de la Parra and A. Córdova, "The border fence and the assault on principles," in Córdova and de la Parra, eds., *A Barrier to Our Shared Environment,* 175–82.

33. J. Knox and D. Markell, "Conclusions," in Markell and Knox, eds., *Greening NAFTA,* 311.

Chapter 17

Transboundary Conservation between the United States and Mexico

New Institutions or a New Collaboration?

Ana Córdova and Carlos A. de la Parra

In a Nutshell ————————————————————————

- The complex management challenges inherent in any transboundary conservation situation are compounded along the lengthy U.S.–Mexico border across which wealth and power disparities are so great.
- The U.S. and Mexican governments have developed a long-standing and varied set of institutional arrangements to manage transboundary natural resources.
- The institutional arrangements fall into three frameworks: centralized collaboration exemplified by the International Boundary and Water Commission/Comisión Internacional de Límites y Aguas, decentralized interagency collaboration deriving from the La Paz Agreement, and truly binational institutions such as the Border Environment Cooperation Commission, the North American Development Bank, and the trinational Commission for Environmental Cooperation.
- The resilience and effectiveness of institutional arrangements are tested under contentious circumstances. An analysis of how the three frameworks have been used (or not) in light of the border wall suggests that as of January 2009, the arrangements have been underutilized.
- Specific ways to revamp and revitalize transboundary collaboration between Mexico and the United States do exist.

Introduction

Wherever political boundaries traverse ecosystems, ecosystem management becomes more complex: information flows are slowed, administrative regimes are fractured, decisions are delayed and often executed with little or no coherence, and responsibilities are masked. The United States and

Mexico jointly share rivers, aquifers, terrestrial ecosystems, and many species that move across one of the longest international boundaries in the world. Although this border appears as a distinct line on the map, the transition of cultures, economies, and ecological conditions is much more gradual and a function of greater interdependence than a map can realistically convey. In addition to the difficulties inherent in managing natural resources in any transboundary context, this case also exhibits great wealth and power disparities between neighboring countries. These disparities are reflected not only in the different institutional, technical, and financial capabilities of the agencies charged with natural-resources management in each country, but also in the degree to which each country can engage its neighbor in cooperative actions and respect for mutual agreements in order to protect the natural environment. The purpose of this chapter is to suggest adjustments in U.S.–Mexico institutional arrangements that may revamp and revitalize conservation collaboration even in the face of changing political imperatives, such as current U.S. concerns with border security.

We first review three evolving frameworks of institutional, transboundary collaboration between the United States and Mexico. Although very valuable informal and formal channels of collaboration exist at state and local government levels as well as at nongovernmental levels, we focus only on formal, federal-level collaboration. We then analyze the effectiveness of these approaches under challenge, using the construction of the international border wall as a test case. Finally, we propose directions in which these approaches might move in order to improve transboundary conservation efforts between the two nations. Our analysis takes into account *(a)* the fact that environmental and natural-resource issues have not historically been priorities in either nation's foreign-affairs agendas, which are instead dominated by trade, drug traffic, immigration, and, more recently, national-security issues, and *(b)* the fact that within each country the agencies mandated with protecting the environment and natural resources tend to be underfunded and have relatively little power in their respective cabinets.

The reflections in this chapter are based on our experiences in Mexico's Secretaría de Medio Ambiente y Recursos Naturales (SEMARNAT, Secretariat of Environment and Natural Resources)—as federal representative from the state of Baja California, secretariat representative to the United States, and ecosystem and land-use planning director at the secretariat's Instituto Nacional de Ecología (National Institute of Ecology)—in addition to our experiences in academic, nongovernmental organization, and state government work on natural-resource management and environmental policy issues in Mexican northern border states. Throughout these experiences, we have interacted extensively with and within the institutional arrangements described here and have observed these processes firsthand.

Discussion

Institutional Frameworks of U.S.–Mexico Environmental Collaboration

The United States and Mexico have more than a century of experience addressing the management of a shared environment. This experience is manifested in at least three types of collaboration, which have been adapted over time to the complexity of natural resources and environmental issues being addressed, changing societal factors, and a common search for better institutional arrangements. In this section, we describe these three types: (1) centralized collaboration represented by the International Boundary and Water Commission/Comisión Internacional de Límites y Aguas (IBWC/ CILA), (2) decentralized collaboration under the auspices of the La Paz Agreement, and (3) binational institutions spawned by the North American Free Trade Agreement (NAFTA). These frameworks have appeared chronologically, with an overall trend toward including a wider range of natural-resource issues, within a broader geographical scope, and with a more integrated, less centrally controlled administrative structure. Despite the evolution toward greater complexity and integration that we describe later, we argue that all three frameworks continue to be useful in collaborative transboundary conservation.

Centralized Collaboration

A precursor to a centralized framework of collaboration on the environment was the 1906 Convention Between the United States of America and the United States of Mexico for the Equitable Distribution of the Waters of the Rio Grande, the first agreement dealing with natural-resource issues signed by the two countries. It settled differences over irrigation waters, and it was created through diplomatic channels. In 1889, the International Boundary Commission emerged to settle territorial disputes between the countries. As transboundary water management became more complex, the International Water Treaty was signed in 1944. This treaty created the IBWC/ CILA to handle bilateral negotiations on water. Environmental quality and integrated water management were not the commission's concerns. The IBWC/CILA mandate was and remains territorial integrity and a politically viable division of waters, centered on a boundary-only/water-only objective (see chapter 1 in this volume).

The new commission was granted the status of an "international body" with all the "privileges and immunities appertaining to diplomatic officers" awarded to its commissioner and staff. As in the International Boundary Commission, in the IBWC/CILA each country has a section working under its own foreign office, but careful emphasis has been placed on specifying that only an engineer can be named commissioner, thereby defining the type of capabilities and discussions the two commissioners would have. Parallel officers are appointed by each government and headquartered in adjacent border towns, working as liaisons on boundary and water-apportionment issues.

IBWC/CILA is a key example of the two federal governments' centralized collaborative response to issues considered highly sensitive but narrow in scope. Nevertheless, it was a departure from the hitherto classic notion of institutions as strictly national entities. In the context of U.S.–Mexico relations, the commission played a pioneering role in moving the discussions and, to some extent, the decision-making power from Washington, D.C., and Mexico City to the boundary line. There is no shortage of complaint along the border over the persistent concentration of power in the national capitals, but IBWC/CILA has been able to incorporate state and local concerns into its efforts, thereby managing the tension between national-sovereignty issues, on the one hand, and everyday interactions and aspirations at local and state levels, on the other. Tasking the two international commissioners with resolving boundary and treaty issues and the operation of water infrastructure was a de facto decentralization of views and ideas, which created the conditions for a new border administrative culture, with two like-minded professionals chosen to lead organizations that would mirror each other. Each was tasked with representing his or her own country, solving issues of shared resources, and maintaining comity with one another.

Decentralized Collaboration

Without the intervention of the foreign ministries—the U.S. Department of State and Mexico's Secretaría de Relaciones Exteriores (SRE, Secretariat of Foreign Relations)—U.S. and Mexican federal natural-resources agencies enjoy rich cross-border collaboration. The framework for this relatively decentralized collaboration was set in 1983, when U.S. president Ronald

Reagan and Mexican president Miguel de la Madrid signed the Agreement Between the United States of America and the United Mexican States on Cooperation for the Protection and Improvement of the Environment in the Border Area, better known as the La Paz Agreement. The border area was defined as 100 kilometers (62 miles) on either side of the political line. La Paz brokered a new model of diplomacy in which environmental agencies outside the national foreign offices would be directly tasked with engaging in cross-border arrangements on issues within their purview. Although the agreement was developed primarily to address pollution, biodiversity conservation has also been pursued under its umbrella.

The La Paz Agreement appointed the U.S. Environmental Protection Agency (EPA) and Mexico's Secretaría de Desarrollo Urbano y Ecología (Secretariat of Urban Development and Ecology, now SEMARNAT)—as its coordinators. Although La Paz was never ratified by the respective national congresses and might be perceived as standing on uncertain legal ground, the executive agreement has stood for more than twenty-five years. It has survived successive administrations in the United States and Mexico and received appropriations from both legislatures. With its emphasis on cooperation "in the field of environmental protection in the border area on the basis of equality, reciprocity and mutual benefit,"[1] La Paz has become a framework under which other cross-border collaborations have taken place.

Under the La Paz framework, natural-resource managers in state and federal agencies along the border share considerable commonality in their work, knowing that the key to successful management of joint resources is to manage them jointly, in both spirit and everyday practice. The unyielding reality of geography requires that agencies establish cooperative conservation programs, information exchanges, seminars, and bilateral commissions with their U.S. and Mexican counterparts. This healthy flow of human energy, information, and even resources includes the U.S.–Mexico environmental program Border 2012; efforts at protecting endangered species and recovering emblematic ones, such as the California Condor (*Gymnogyps californianus*) project headed jointly by the U.S. Fish and Wildlife Service and Mexico's National Ecology Institute; and the sister parks agreements between SEMARNAT and the U.S. Department of the Interior. Even agencies from the ten border states have cooperated together or pooled resources with the federal governments. But the La Paz Agreement and the particular arrangements operating under its umbrella are workable only when its key words *cooperation*, *reciprocity*, and *mutual benefit* are honored.

Binational Institutions

A third framework for transboundary collaboration appeared in 1993, ten years after La Paz and fifty years after IBWC/CILA, with NAFTA and its environmental side agreement (the North American Agreement on Environmental Cooperation). Unlike the previous two arrangements, NAFTA created truly binational institutions: the Border Environment Cooperation Commission (BECC) and the North American Development Bank (NADBank), staffed by Mexicans and Americans but representing neither country; and a trinational institution, including Canada, the Commission for Environmental Cooperation (CEC), staffed by nationals of all three countries.

The CEC includes within its continental and comprehensive mandate the conservation of North America's wildlife, plants, and habitats, particularly in natural protected areas (see chapter 16 in this volume). BECC and NADBank have a narrower geographic scope on either side of the international boundary—100 kilometers (62 miles) on the U.S. side and 300 kilometers (186 miles) into Mexico. These two institutions were created to solve the pressing problem of insufficient environmental infrastructure along the border. BECC's role is to review and certify projects for their conformity to a set of criteria that would render them more socially and environmentally sustainable. Once a project advances beyond the process of certification by BECC, funding becomes available from NADBank. Both agencies operate on an annual budget that is met in equal shares by both countries.

A distinguishing feature of binational collaboration through these sister agencies is that they are staffed by both Mexicans and Americans within the same organization. Unlike the IBWC/CILA model of separate U.S. and Mexican sections responding to their own foreign-affairs secretariats, BECC and NADBank operate with a binational staff that shares a single office building, sits around the same table, analyzes the same data, tackles the same problems, reaches decisions jointly, and, since 2006, shares a single board of directors. Leadership of the agencies alternates between the two countries, with an adjunct leader appointed from the partner country. Binational teams of administrators and technical personnel form under each agency administration, serving to erase the sense of borders within the organization.

A drawback of BECC and NADBank's setting is their isolation. Although the institutions are seated at the border, their boards of directors are composed of bureaucrats in Washington, D.C., and Mexico City. BECC and NADBank executives must therefore work doubly hard to convey their views and navigate through the political mazes of both national governments.

A Challenge to Binational, Transboundary Collaboration

As has been described, there is a long-standing, varied, and evolving web of transboundary, collaborative conservation efforts between the United States and Mexico at the federal level. However, these conventions, agreements, memoranda of understanding, and cross-agency programs are not uniformly effective or functional. There is wide variation in vitality, functionality, and effectiveness in them. Some institutional arrangements exist mostly on paper, whereas others show extensive collaboration. As discussed in other chapters of this book, the strength of the collaboration sometimes depends on the specific agency officials managing the collaborative activities within these bodies. As a consequence, the vitality of the relationship may be bound to administrative periods.

The litmus test of any agreement or collaborative arrangement comes when one of the parties acts unilaterally against the spirit or word of the agreement. The robustness and resilience of collaboration are tested under such circumstances, in which often local stakeholders are in stark disagreement with their distant central governments. There have certainly been many such occurrences in the course of U.S.–Mexico transboundary conservation efforts, but the recent emblematic case has been the construction of a border wall along the U.S.–Mexico border without consultation, discussion, or analysis of its environmental, social, and economic impacts on either side of the border.

When in 2006 the U.S. Congress gave the order to the Department of Homeland Security to secure the border, the wide-ranging waiver it came with responded to internal political needs (see chapter 15 in this volume). The order became an assault on the principles espoused by U.S. society for more than a century, as exhibited by the thirty-plus pieces of legislation waived by former secretary of homeland security Michael Chertoff. It may also be an assault on the U.S. Constitution, as suggested by the organizations that have filed lawsuits in the U.S. Supreme Court.

In addition to violating the values expressed in the waived legislation, this action also transgressed the values and principles of bilateral agreements with Mexico. Under La Paz, both nations "agree to cooperate in the field of environmental protection in the border area on the basis of equality, reciprocity, and mutual benefit," and "[t]he Parties shall assess, as appropriate in accordance with their respective national laws, regulations and policies, projects that have significant impacts on the environment of the border area, that appropriate measures may be considered to avoid or mitigate adverse environmental effects."[2]

In building the border wall, the United States is perceived by many as failing to honor its commitments under the La Paz Agreement. This U.S. attitude is a dramatic departure from its position in the mid-1980s, when untreated sewage from Tijuana began flowing into San Diego. Roberto Sánchez-Rodríguez argues that under the Reagan administration, the United States was rigid and inflexible as it demanded that Mexico honor La Paz and forced it to undertake environmental infrastructure investments that were extremely costly and beyond its investment capabilities at the time.[3] The U.S. government has thus seemingly upheld La Paz principles when Mexico's actions have affected its environment, but not always when its own actions affect Mexico's environment. As much as we recognize that in many examples worldwide it is common for more powerful parties to uphold agreements at their convenience, actions undertaken on the grounds of the two countries' disparities in economic and political power will ill serve the principles put in place by both nations to preserve natural resources for generations to come.

This unilateral action on the part of the United States, with broad foreseeable environmental impacts, and the U.S. disregard for La Paz prompted a reaction from the Mexican Senate,[4] which issued a request that SEMARNAT and SRE engage their U.S. counterparts to jointly assess potential environmental impacts of the wall. At the request of SEMARNAT secretary Juan Elvira, several scientific workshops were subsequently held with scholars and practitioners from both sides of the border to assess impacts, their economic costs, and the various fora and mechanisms through which a constructive binational dialogue on the wall might be conducted.[5]

Mexico's SRE has sent as many as five diplomatic notes on the issue to the United States but has received no formal diplomatic response from its neighbor. Mexican presidents Vicente Fox and Felipe Calderón[6] have publicly chastised the U.S. government regarding the wall, but the diplomatic conversation has otherwise remained relatively quiet.[7] Secretary Elvira had several informal conversations with former U.S. interior secretary Dirk Kempthorne about the results of the workshops and even a brief exchange with former homeland security secretary Michael Chertoff. Given the complexities of U.S.–Mexican relations, however, Mexico's official reaction, up until the end of the Bush administration, has been quite reserved—bewildering observers and even some U.S. government officials.

Searching for a More Effective Collaboration on Conservation Issues

It is apparent from the previous discussion that in addition to direct contact between the foreign ministries of each country, Mexico and the United States have a variety of formal institutional channels to address environmental issues of joint concern. Those channels have been evolving toward greater complexity and integration according to both parties' changing needs. We have suggested that until now, binational communication and action on the border wall issue has been less than optimal. We now propose ways that discussion and action through existing institutions, using existing frameworks, can be strengthened and revitalized to give rise to a new collaboration.

Using the Centralized Framework

The formal diplomatic dialogue between the U.S. State Department and Mexico's SRE on natural resources has been exercised mainly through IBWC/CILA channels. So far, however, IBWC/CILA has been minimally involved in decisions involving the construction of the wall (beyond modifications Homeland Security made to levee design in the lower Rio Grande to conform to IBWC specifications). Perhaps this is because IBWC/CILA's mandate differs from the issues the wall is meant to address—undocumented immigration, drug trafficking, and national security—and because Homeland Security did not significantly involve other agencies in wall-construction decisions. There are, however, documented instances in which border barrier construction occurred on Mexican territory, in places such as Agua Prieta[8] and Nogales,[9] Sonora, and in rural sites observed by wildlife biologists. To the extent that the border wall must respect the official international boundary and can modify hydrological flows, the IBWC/CILA should be adequately consulted in its design and construction.

Moreover, the U.S. State Department and SRE, as the focal points for cross-border interaction in this framework, have so far failed to set up substantive exchanges of information that might have produced recommendations to optimize both ecological soundness and improved security. This failure seems in part due to the State Department's willingness during the Bush administration to be displaced by Homeland Security on U.S.–Mexican border issues. This displacement may have made the State Department a less-relevant counterpart for SRE in bringing the wall and related border environmental issues to the table for substantive discussion.

Using the Decentralized Framework

Despite various decentralized agency interactions taking place during 2008 between SEMARNAT and the U.S. Department of the Interior, from technical staff meetings to the encounters between secretaries, no formal agreement or plan of action was adopted to assess jointly the impacts of the border wall on wildlife and ecosystems, much less to implement mitigating actions. Given these two agencies' mandate to protect species, ecosystems, and natural processes in the border region and elsewhere, we suggest that continued efforts be made to formalize ways to address jointly the threats the wall poses to the natural environment. From conversations with Interior Department and SEMARNAT technical staff, we know that both agencies received indications from the State Department and SRE at different moments not to raise the wall's profile beyond conversations. But if the La Paz Agreement is the exemplar framework for cross-border agency collaboration on environmental issues, it can be argued that the national coordinators of the agreement, the U.S. EPA and SEMARNAT, can and should operate outside of the foreign ministries' direct lead. Some may argue that La Paz covers only the EPA's mandate—pollution prevention and not natural-resources conservation—yet this restriction is not warranted by the agreement's text and the history of its implementation or from the perspective of Mexico's implementing agency, SEMARNAT. The fact that the Interior Department was involved in the Border XXI Program under La Paz, when one of the working groups covered natural resources, supports the interpretation that even in the United States, La Paz was understood to extend beyond the EPA's administrative jurisdiction. In addition, La Paz has explicit provisions (Art. 9) for the inclusion of other agencies when necessary for the purposes of the agreement.

Finally, a new annex to the La Paz Agreement might be developed on the border wall issue or on the issue of transboundary conservation cooperation in more general terms, based on Article 3, which reads: "Pursuant to this Agreement, the Parties may conclude specific arrangements for the solution of common problems in the border area, which may be annexed thereto."

There is no technical reason why issues such as the border wall or broader transboundary conservation collaboration cannot be discussed under the La Paz framework or why the Interior Department cannot interact directly with SEMARNAT under this umbrella. Further, because the U.S. Department of Agriculture, through its Forest and Natural Resources

Conservation Services, also manages significant natural resources with transboundary impact, including this agency in such interactions would be useful.

Using the Binational Institutions Framework

Of the NAFTA-related institutions, only the CEC, which covers all of North America, has an explicit mandate to address transboundary conservation issues, including those affected by the border wall (see chapter 16 in this volume). The CEC would likely analyze the potential participation of Canada in such activities, depending on *(a)* the extent to which transboundary conservation impacts on the U.S.–Mexico border will affect species or ecosystems of concern to conservation interests in Canada and *(b)* whether the issues occurring today on the Mexican border with the United States might someday also occur on or provide lessons for the Canadian border with the United States. In any event, we believe it would be very relevant and useful for this issue to be taken up in this forum as well. In contrast, BECC and NADBank, with their mandate more focused on the U.S.–Mexico border zone, have already become involved in certifying and funding water conservation in the Río Conchos basin in northern Mexico. It might be worth exploring future funding from these two agencies for wetlands restoration and other transboundary, collaborative conservation projects.

Conclusions

A wealth of strategies and institutions exist through which Mexico and the United States can seek constructive collaboration—involving dialogue, analysis, and mitigation—regarding transboundary conservation issues, including the border wall. However, these arrangements have so far not been exploited to their full potential. The deficiencies that we have described in collaborations across the border are also present in collaborations among federal agencies within each country. In the construction of the border wall, the U.S. Department of Homeland Security has been working largely without collaboration from, input from, or substantive consultation with the Departments of Interior, Agriculture, and State. In other binational transboundary conservation issues, such as the impacts of lining the All-American Canal on the Andrade Mesa wetlands in Mexico, lack of federal interagency coordination within each country has led to less-than-optimal results for the environment. In addition to the ways we have outlined for revitalizing transboundary conservation, improved interagency collaboration *within* each country is paramount for the benefit of the shared environment and of both nations.

The analysis presented thus far pertains to the period up until January 2009. There are grounds for optimism in anticipating the policies of the new U.S. federal administration. President Barack Obama consistently emphasizes the importance of upholding U.S. values and ideals in matters of security, "rejecting the false choice between [our] safety and our ideals,"[10] "because living our values doesn't make us weaker. It makes us safer and it makes us stronger."[11] We are hopeful that the U.S. ideals upheld will include the use of existing legislation to protect future generations from an impoverished environment. The fact that the new secretary of homeland security, Janet Napolitano, is knowledgeable of border issues and aware of the multiple implications of the wall raises additional prospects for security policy based on a broader vision that recognizes environmental as well as national-security needs. We are also hopeful that the U.S. ideals of upholding the rule of law and international agreements can again be bases upon which to build a better relationship with its neighbor to the south. Secretary of State Hillary Rodham Clinton has stated "robust diplomacy and effective development are the best long-term tools for securing America's future."[12] If diplomacy is one of the best long-term tools for securing the future of the United States, then it may be possible to revisit the wall issue and perhaps redefine it.

The degree to which the various frameworks of binational collaboration are strengthened, as discussed in this chapter, will impact a host of issues related to natural resources and the environment. The new U.S. federal administration signals a valuable opportunity for revitalized transboundary collaboration on conservation between Mexico and the United States. This is an opportunity that neither country can afford to miss.

Acknowledgments

We thank the editors and anonymous reviewers for very useful feedback provided during the writing of this chapter.

Notes

1. "Agreement Between the United States of America and the United Mexican States on Cooperation for the Protection and Improvement of the Environment in the Border Area" (La Paz Agreement), signed August 14, 1983, Art. 1, full text available at http://www.epa.gov/usmexicoborder/docs/LaPazAgreement.pdf.

2. Ibid., Art. 1 and 7.

3. R. Sánchez-Rodríguez, *El medio ambiente como fuente de conflicto en la relación binacional México–Estados Unidos* (Tijuana, Mexico: El Colegio de la Frontera Norte, 1990).

4. See the Mexican Senate's *Diary of Debates*, November 14, 2006, and March 1, 2007: Senado de la República, *Diario de Debates*, LX Legislatura, available at http://www.senado.gob.mx/diario, accessed February 19, 2009.

5. For the results of the first workshop, see A. Córdova and C. A. de la Parra, eds., *A Barrier to Our Shared Environment: The Border Fence Between the United States and Mexico* (Mexico City and San Diego: Secretaría de Medio Ambiente y Recursos Naturales, Instituto Nacional de Ecología, El Colegio de la Frontera Norte, and Southwest Consortium for Environmental Research and Policy, 2007), available at http://www.ine.gob.mx.

6. See, for instance, J. C. McKinley, "Mexican president assails U.S. measures on migrants," *New York Times*, September 3, 2007, available at http://www.nytimes.com/2007/09/03/world/americas/03mexico.html.

7. Our analysis covers events up until January 2009.

8. See J. A. Roman, "Confirma Washington incursiones en México para alzar el muro fronterizo," *La Jornada*, March 6, 2007, available at http://www.jornada.unam.mx/2007/03/06/index.php?section=politica&article=009n1pol.

9. Claudia Gil, urban planning director for the City of Nogales, presentation at the Instituto Nacional de Ecología/El Colegio de la Frontera Norte/Arizona State University 2008 Workshop on Environmental Impacts of the Border Wall.

10. B. H. Obama, Presidential Inaugural Address, January 20, 2009, available at http://www.whitehouse.gov/the_press_office/President_Barack_Obamas_Inaugural_Address/, accessed April 20, 2009.

11. B. H. Obama, Address to Joint Session of Congress, February 24, 2009, available at http://www.whitehouse.gov/the_press_office/Remarks-of-President-Barack-Obama-Address-to-Joint-Session-of-Congress/, accessed April 20, 2009.

12. H. R. Clinton, speech at the U.S. Department of State, January 22, 2009, available at http://www.state.gov/secretary/rm/2009a/01/115262.htm, accessed April 20, 2009.

Guiding Principles for Successful Transboundary Conservation

This conclusion reflects on the volume's most significant insights and organizes them into five guiding principles for structuring efforts and policies to conserve the shared environment of Mexico and the United States. In offering these principles, we suggest that they are enduring and applicable to other transboundary situations. Then, in a follow-up, "Energizing Transboundary Conservation—Three Easy Steps," we provide three simple, feasible policy recommendations that can be rapidly and easily implemented today.

In brief, the most significant insights for framing transboundary conservation to emerge from the chapters are that (1) the environment shared by Mexico and the United States extends far beyond the political line, as may the causes of, consequences of, and solutions to transboundary conservation challenges; (2) successful transboundary conservation happens at multiple scales, from the local to the federal, and should include diverse stakeholders; (3) informal collaborative and nongovernmental processes are instrumental in promoting transboundary conservation; (4) formal binational and governmental institutions are likewise crucial to establishing binational responsibilities for protecting shared environments and facilitating transboundary conservation; and (5) the concept of ecosystem services is useful for framing transboundary conservation in terms of mutual interests between countries.

Geographic Scope

One of the main lessons of this book is that conservationists and decision makers would be well served by thinking beyond the border region. Many studies of transboundary conservation have focused on the narrow border region,[1] perhaps because the La Paz Agreement and the North American Free Trade Agreement (NAFTA) and its environmental side provision defined the U.S.–Mexico border region as a 100-kilometer (62-mile) fringe on either side of the political line. Other scholars of transboundary conservation, although not limiting themselves to the narrow border region, have discussed conservation in terms of ecosystems, ecoregions, or biomes.[2]

In contrast to the narrower approaches, the chapters in this volume demonstrate that the breadth and depth of the ecological connections between the United States and Mexico may extend beyond the border region or even beyond shared biomes. For example, in chapter 7, Chester and McGovern point out how land-use change and habitat destruction in Central America and southern Mexico influence the diversity of migratory songbirds enjoyed by bird-watchers in the United States and Canada. Likewise, in chapter 11, Medellín demonstrates how the destruction of bat roosts in the U.S.–Mexico borderlands region can diminish the pollination services by bats that support tequila production in central Mexico. Calderon-Aguilera and Flessa argue in chapter 10 that fisheries in the Gulf of California are impacted by water diversions in the Colorado River hundreds of miles to the north.

Even chapters that focus on habitats much closer to the physical border speak to broader geographical connections. Culver and her colleagues' explanation in chapter 6 of the drivers of habitat fragmentation and Sharp and Gimblett's examination in chapter 14 of the physical impacts of migration and smuggling in borderland protected areas demonstrate connections to national-security policies and social drivers originating elsewhere.

We suggest that decision makers think well beyond the border when shaping transboundary environmental policy. In chapter 9, López-Hoffman, Varady, and Balvanera suggest that the concept of ecosystem services may help us think expansively about cross-border connections by elucidating how drivers of environmental change in one country can affect a wide range of ecosystem services and human welfare in another.

Multiple Scales and Diverse Stakeholders

Transboundary collaborations happen at multiple scales, from the local to the federal, and should include diverse stakeholders: the transboundary conservation efforts described in this book range from local projects initiated by local stakeholders to formal binational diplomacy between governments. Chapters 2, 3, 4, 8, and 13 demonstrate how local stakeholders from both sides of the border have worked together to conserve a shared environment or species. Briggs and his coauthors (chapter 3) describe how local stakeholders and nongovernmental organizations (NGOs) are promoting the conservation and restoration of border wetlands on the Rio Grande/Río

Bravo. Piekielek (chapter 13) discusses how park managers from El Pinacate in Sonora, Mexico, and Organ Pipe in Arizona actively work together as "sister parks" despite obstacles ranging from federal prohibitions on foreign travel to funding shortages. Zamora-Arroyo and Flessa (chapter 2) report that both U.S. and Mexican NGOs and local stakeholders of the Colorado River delta have been negotiating directly with state governments and water districts in the two countries to obtain water rights for the support of the delta. As Starks and Quijada-Mascareñas (chapter 4) explain, the governments of Native nations in the United States and Indigenous citizens in Mexico are actively pursuing transboundary conservation partnerships and should be further incorporated as key decision makers and participants in regional conservation and planning processes. These examples indicate that federal-level action is not necessarily a precursor to local or regional transboundary conservation.

Informal Processes

Transboundary conservation can happen through informal, collaborative, and nongovernmental processes. Although scholars of international environmental cooperation recognize the role of civil society and nongovernmental actors, most have emphasized formal intergovernmental processes as promoting transboundary conservation (discussed under the next heading).[3] The contributors to this book have illustrated how a range of actors, including NGOs, civil society groups, and academics, have been successful in promoting collaborative transboundary conservation projects. Ceballos and his coauthors as well as Calderon-Aguilera and Flessa illustrate academic researchers' role in promoting conservation in the transboundary grasslands of Chihuahua and New Mexico and in the upper Gulf of California, respectively. Chapters 3 and 7 emphasize the importance of environmental NGOs and other informal processes in facilitating conservation across the border. In the former chapter in particular, Briggs and his coauthors illustrate that NGOs have been successful in convening, facilitating, and funding cooperative binational restoration in Big Bend and the Forgotten Reach of the Rio Grande/Río Bravo.

It is important to note that in some cases, processes that begin as informal collaborations between nongovernmental stakeholders can become formal intergovernmental policy. Zamora-Arroyo and Flessa show in chapter 2 that NGOs and water district managers in both countries are now working together in a formal intergovernmental forum known as the Colorado

River Joint Cooperative Process. The authors suggest that an outcome of the process might be a new minute to the International Water Treaty of 1944, specifically allocating water to maintain wildlife and habitat in the Colorado River delta.

One of the most surprising ideas to emerge from the chapters is the notion that NGOs and local stakeholders have been using crises to create spaces for binational collaboration. According to Zamora-Arroyo and Flessa, the ongoing drought in the western United States has opened the possibility of dedicating instream flows of water for the Colorado River delta. Prior to the drought, conservationists and water managers were at loggerheads, and it seemed improbable that either the Mexican government or the U.S. government would dedicate valuable water to the delta. The drought forced both groups to develop creative solutions for dealing with water shortages in the Colorado River basin; among these solutions are suggestions for acquiring alternative water sources. Chapters 13, 15, and 17 echo the theme that crises create space for conservation, noting how border security and the wall may be renewing interest in collaborative, binational conservation in the borderlands.

Binational Institutions and Treaties

The aforementioned examples demonstrate that academia, NGOs, and civil society, working collaboratively across borders, can break down barriers between countries to promote creative solutions for managing shared environments. Interagency collaborations involving researchers and land managers are also key. Nonetheless, the authors also call for strengthening existing formal intergovernmental institutions and creating new frameworks and treaties to protect and conserve the environment shared by the United States and Mexico. Zamora-Arroyo and Flessa argue in chapter 2 that the 1944 treaty between the United States and Mexico to allocate Colorado River water needs to be amended to allocate water for the river's delta as well. Similarly, López-Hoffman and her coauthors stress in chapter 9 the need for new international agreements to collaboratively manage and equitably distribute water from transboundary aquifers. In chapter 16, Mumme and his coauthors argue that the North American Commission for Environmental Cooperation, established by a side agreement to NAFTA in 1994, occupies a unique and valuable niche in North American relations that is particularly vital for transboundary conservation. Because biodiversity conservation across the U.S.–Mexico border presents an administrative challenge that exceeds the capacity of either national government, these authors believe that formal binational forums such as the Commission for Environmental Cooperation are paramount for addressing the vast ecological linkages between the countries of North America.

An Ecosystems Services Framework

Ecosystem services can frame transboundary conservation. Several authors in this book have suggested a novel approach to transboundary conservation by framing it in terms of shared ecosystem services. Chapters 9, 11, and 12 point out how the many species that range or fly across the U.S.–Mexico border provide services such as pollination and pest control for crops. Further, they demonstrate how drivers of ecosystem change in one country may affect the delivery and quality of ecosystem services and consequently human well-being in another country. In chapter 2, Calderon-Aguilera and Flessa show how efforts to increase water provisioning in the United States are altering ecosystems and ecosystem services in Mexico, in turn impairing human well-being. López-Hoffman and her coauthors suggest that ecosystem services can be used to frame transboundary conservation as being in the two countries' mutual interest. They point out that actions taken in this mutual interest create incentives for working together rather than in contradiction to one another.

How Does Successful Transboundary Conservation Happen?

The chapters in this volume have demonstrated that the United States and Mexico share a common environment, illustrating how they are united by the water that flows across their border in rivers and under the border in aquifers, by the species that fly across the border and the animals that range across the border, and by the ecosystems that bestow vital services to people in both countries. The chapters have also shown that the species, spaces, and ecosystem services that unite the United States and Mexico are threatened by the common drivers of land conversion, habitat degradation and fragmentation, and overconsumption of water, as well as border security, undocumented migration, and drug trafficking. These driving forces, as illustrated in chapters 1 and 5, have been caused by millennia of natural processes and human actions. Because these drivers continue to shape the environment, effective conservation strategies should strive to adapt to and cope with them. Just as the United States and Mexico share an environment, they also have joint responsibility for its conservation and protection.

This book's goal has been to suggest binational, cooperative strategies for conserving the common environment of the United States and Mexico. Throughout the pages of this book, the authors have sought answers to the question, How does successful transboundary conservation happen? The efforts they describe suggest that successful transboundary conservation can happen when conservationists and decision makers focus beyond the line—and beyond the borderlands—at critical bird stopover points throughout the Americas or within the vast Colorado River system far to the north of its delta in Mexico or on the ecological processes in Mexico that support life-sustaining ecosystem services in the United States and vice versa. The authors illustrate how transboundary conservation can happen at many scales, through both governmental and nongovernmental processes. At the same time, they emphasize how support from effective binational institutions capable of addressing the vast ecological linkages between the countries is critical. They suggest that successful transboundary conservation efforts can sometimes arise through crises—when stakeholders are prepared to transform crises into opportunities for positive change. Above all, they demonstrate how successful transboundary conservation happens when the decision-making sphere includes not only legislators and government agency headquarters in Mexico City and Washington, D.C., but also tribes, NGOs, land managers, researchers working on the ground, binational conservation alliances, and many institutions and people in between.

Notes

1. K. Hoffman, ed., *The U.S.–Mexican Border Environment: Transboundary Ecosystem Management*, Southwest Consortium for Environmental Research and Policy Monograph Series, vol. 15 (San Diego: San Diego State University, 2006).

2. C. Chester, "Civil society, international regimes, and the protection of trans-boundary ecosystems: Defining the International Sonoran Desert Alliance and the Yellowstone to Yukon Conservation Initiative," *Journal of International Wildlife Law and Policy* 2(2) (1999): 159–203; C. Chester, *Conservation across Borders: Biodiversity in an Interdependent World* (Washington, D.C.: Island Press, 2006).

3. Chester, "Civil society, international regimes, and the protection of transboundary ecosystems"; L. Susskind, W. Moomaw, and K. Gallagher, eds., *Transboundary Environmental Negotiation: New Approaches to Global Cooperation* (San Francisco: Jossey-Bass, 2002).

Energizing Transboundary Conservation— Three Easy Steps

Where to begin? Throughout this volume, the authors have described the breadth of the environmental connections between the United States and Mexico as well as the ecological, social, and political complexity of transboundary conservation. Going further, they have suggested numerous ways of improving binational management in the shared environment. Reflecting on the guiding principles that have emerged in this book, the editors offer three simple, feasible policy recommendations. Because these ideas follow on existing binational institutional arrangements between the countries, they are therefore actionable in the short term.

First, under the auspices of the 1983 La Paz Agreement, the U.S. and Mexican governments should form a binational task force to monitor the environmental impacts of border security. In La Paz, the nations agreed to cooperate in environmental protection in the border area and to "assess ... projects that have significant impacts on the environment of the border area, that appropriate measures may be considered to avoid or mitigate adverse environmental effects."[1]

The task force should be composed of diverse stakeholders from both countries, including local, state, and federal land-management agencies; nongovernmental organizations (NGOs); academic researchers; and the local and Indigenous peoples most affected by security activities. It should begin by developing and implementing protocols for monitoring environmental and social impacts of border security. After a reasonable period of monitoring, the working group should then set priorities for and oversee the implementation of mitigation efforts. The binational effort will not only lessen the ecological effects of security infrastructure and activities on borderland landscapes and species, but will reawaken a spirit of transboundary conservation cooperation in the wake of unilateral security actions.

Second, we suggest that in the spirit of Minute 306 of the 1944 International Water Treaty between the United States and Mexico, the countries should support restoration of riparian, wetland, and estuarine habitats such as those of the Colorado River and the Rio Grande/Río Bravo and their deltas. The two governments should make funds available to enable the International Boundary and Water Commission/Comisión Internacional de Límites y Aguas, with the advice of binational technical task forces, to support NGOs, community groups, universities, and the private sector. These groups would help restore riparian and delta habitats affected by surface-water allocations, groundwater utilization, and water policies and practices. Such studies and activities would define the ecosystem functions desired by stakeholders in each country and allocate water to restore those functions. Funds should support projects that are binational and collaborative and that address the greatest environmental needs, regardless of location.

Third, the Commission for Environmental Cooperation (CEC) should be reinvigorated. The CEC was created by the North American Agreement on Environmental Cooperation (NAAEC), the environmental side agreement to the North American Free Trade Agreement. Of all the binational arrangements for transboundary conservation, the CEC is the only one with a geographic mandate broad enough to address the vast ecological linkages between the countries and a topical mandate sufficiently expansive to consider the spectrum of environmental and social changes shaping the U.S. and Mexico's shared environment. The CEC's three principal functions—scaling ecological linkages across North America, organizing networks of stakeholders, and spotlighting issues of transboundary environmental concern through investigations—suit it to work with stakeholders to address transboundary ecological and conservation concerns (see chapter 16 in this volume).

The CEC's Transboundary Environmental Impact Assessment mechanism, envisioned in the NAAEC enabling document, can add immediate force to protecting the environment of one country against unintended impacts of actions and policies in the other country. We urge the two countries, after fifteen years of nonaction, to negotiate and implement a workable and forceful environmental impact agreement for conserving the U.S. and Mexico's shared environment.

Since the CEC's creation in 1994, the United States and Mexico have generally supported its mission. Nevertheless, it has been underfunded and is seen as weak. Because it has lacked the resources and thus the "teeth" to identify problems, investigate them, and enforce corrective measures, it has yet to fully realize its promising mandate.

For the CEC to be effective requires two important changes: (1) a redesign to make it more inclusive of affected communities and other stakeholder groups, and (2) far greater and longer-term financial commitment by the participating countries.

A redesigned CEC would be uniquely situated to meet transboundary conservation principles set forth in this book: thinking beyond the border to continentwide ecological connections; incorporating diverse stakeholders; encouraging cross-border collaboration at many scales, from local to federal; and framing conservation in terms of ecosystem services. We believe this moment to be ideal for the United States and Mexico to work together to reinvigorate the CEC.

At a time of heightened environmental changes—when growth is depleting water resources and stressing transboundary ecosystems, when border security activities are transforming borderland landscapes and preventing species movements across the border, when actions and policies in one country are affecting ecosystem services and human well-being in the other country, and when the two governments can start afresh and bring new perspectives to thorny problems—Mexico and the United States should seize the opportunity to work together in conserving their shared environment. They would be wise to begin with three simple, feasible, and actionable steps: develop a binational task force on border security, restore transboundary riparian ecosystems, and reinvigorate the CEC.

Note

1. "Agreement Between the United States of America and the United Mexican States on Cooperation for the Protection and Improvement of the Environment in the Border Area" (La Paz Agreement), signed on August 14, 1983, Art. 1, full text available at http://www.epa.gov/usmexicoborder/docs/LaPazAgreement.pdf/.

Acknowledgments

This book is the result of ongoing conversations between the volume editors and the community of stakeholders working on U.S.–Mexico transboundary conservation. The origin of the book can be traced to a symposium about transboundary conservation that Laura López-Hoffman and Karl Flessa organized at the 2006 Meeting of the Society for Conservation Biology. After the meeting, Rodrigo Medellín and Gerardo Ceballos urged Flessa and López-Hoffman to develop a book on the topic; they recognized the value of a volume to address the social, political, and ecological challenges of conserving biodiversity in the environment shared by the United States and Mexico.

This volume found a home within the University of Arizona Press's new book series THE EDGE: Environmental Science, Law, and Policy. Our effort was greatly energized by lead series editor Marc Miller's vision and passion for bridging science and policymaking. Along with series coeditors Jonathan Overpeck and Barbara Morehouse, Miller developed the idea of a new book series to translate science into workable policy options. Our book's inclusion in it helps us reach both policy and research communities. Dean Toni Massaro at the James E. Rogers College of Law provided enthusiastic support for the idea of THE EDGE series, and both financial and institutional support for this volume.

Coordinating this volume has been a great pleasure from the start, thanks to the support provided by the authors. Their enthusiasm, creativity, hard work, and patience through multiple revisions has made this effort a joy for us as editors. We particularly appreciate the authors' participation in a workshop in late May 2008 at the University of Arizona's Biosphere 2. This meeting laid the groundwork for focused, policy-oriented chapters and allowed the authors, volume editors, and series editors collectively to refine the book's key themes and purposes through interactive, in-person discussions. The gathering further established the sense of shared investment in transboundary conservation between many of its practitioners from both countries.

This volume would not have been possible without the support of the University of Arizona Press. We thank Christine Szuter, the press's former director, who recognized this work's early potential and strongly encouraged it. We also thank Allyson Carter, editor in chief, for patiently guiding us through the publication process. We appreciate Tammie Brown's persistence in working diligently to help procure financial support for the series and this volume. Thanks also go to Natasha Varner, Barbara Yarrow, Nancy Arora, Annie Barva, and Linda Gregonis, who provided invaluable help with the book's artwork and production.

We thank Mickey Reed for skillful production of the maps that appear throughout the volume.

We are deeply grateful to the Udall Center for Studies in Public Policy for its staff and facilities' support. Stephen Cornell and Robert Merideth provided important counsel. We thank Anne Browning-Aiken for her skillful facilitation at the authors' meeting. Denise Lum helped with formatting the manuscripts. Ian Record's expertise on Native nations was very valuable. And we are especially grateful to designer Renee La Roi for her creativity in developing the beautiful cover artwork.

Without Rocío Brambila's logistical support, the authors' workshop would not have been possible. We are also grateful to Travis Huxman for hosting the workshop at Biosphere 2 and to Val Kelly for her help during the meeting. Thomas Sheridan offered valuable advice and early constructive critique of the project. Michael Crimmins provided expertise on North American climatological patterns. Research assistants Matt Skroch and Crystal Whalen helped develop the "In a Nutshell" chapter summaries. Lisa Graumlich and Graciela Schneier-Madanes provided encouragement throughout the project. And the anonymous reviewers provided many insights that made this book stronger.

Finally, funding from the University of Arizona's James Rogers College of Law, the Institute of the Environment (formerly the Institute for the Study of Planet Earth), the Biosphere 2 Institute, the Udall Center, the Research Coordination Network–Colorado River Delta, and the Morris K. Udall Foundation made this effort possible.

About the Contributors

Patricia Balvanera studies links among biodiversity, ecosystem function, and associated services to society. She was awarded the Man and the Biosphere–UNESCO award for Young Scientists and is an Aldo Leopold Leadership Fellow.

José F. Bernal Stoopen is general director of Wildlife and Zoos in Mexico City. He holds degrees from Texas A&M University and Universidad Autónoma Metropolitana–Xohimilco.

Mark Briggs is with the World Wildlife Fund's Chihuahuan Desert Program and is currently focusing on implementation of binational river rehabilitation along the Rio Grande.

Luis E. Calderon-Aguilera is professor at the Centro de Investigación Científica y de Educación Superior, Ensenada, Baja California, and is interested in ecosystem-based management and global change. He has worked on fisheries ecology in the upper Gulf of California since 1984.

Gerardo Ceballos is an ecology and conservation biology professor at the Universidad Nacional Autónoma de México working on endangered species, nature reserves, and land planning.

Charles C. Chester is author of *Conservation across Borders: Biodiversity in an Interdependent World* (2006) and teaches at Brandeis University. He serves on the board of the Yellowstone to Yukon Conservation Initiative.

Ana Córdova is research professor at El Colegio de la Frontera Norte in Ciudad Juárez. Formerly director general of Landscape Ecological Land-Use Planning and Ecosystem Conservation at the Instituto Nacional de Ecología of the Secretaría de Medio Ambiente y Recursos Naturales, her prior experience includes consultancies with the Commission for Environmental Cooperation and The Nature Conservancy.

Melanie Culver is assistant professor of wildlife and fisheries conservation and management and part of the Arizona Cooperative Fish and Wildlife Research Unit, U.S. Geological Survey, in the School of Natural Resources and Environment at the University of Arizona. Her research involves conservation genetics for mammals, birds, fish, reptiles, amphibians, and invertebrates.

Ana D. Davidson is a U.S. National Science Foundation research fellow at the Instituto de Ecología, Universidad Nacional Autónoma de México. Her research focuses on grassland ecology and biodiversity conservation.

Carlos A. de la Parra is a professor at El Colegio de la Frontera Norte in Tijuana, Baja California, studying U.S.–Mexico relations, bilateral environmental management, and sustainability. He has previously served as the university's provost as well as regional officer for Mexico's Secretaría de Medio Ambiente y Recursos Naturales in Baja California and as the secretariat's representative in the Mexican Embassy in Washington, D.C.

Karl W. Flessa is professor and head of the Department of Geosciences at the University of Arizona. A paleontologist by training, he has worked on the conservation biology and restoration ecology of the Colorado River delta since 1992.

Ed L. Fredrickson is a scientist working with the U.S. Department of Agriculture–Agricultural Research Service Jornada Experimental Range.

Osiris Gaona is research specialist at the Instituto de Ecología of the Universidad Nacional Autónoma de México, where she focuses on conservation of bats in Mexico and North America, including work with the Commission for Environmental Cooperation Trilateral Committee.

Randy Gimblett is professor in the School of Natural Resources and Environment at the University of Arizona. His research includes spatial dynamic ecosystem modeling, environmental perception, and human movement in visitor landscapes.

Louis A. Harveson is professor of wildlife management at Sul Ross State University and director of the Borderlands Research Institute for Natural Resource Management.

Patricia Moody Harveson is assistant professor of conservation biology at Sul Ross State University. Her research includes carnivore ecology, desert ecosystems, and the use of modeling in the conservation of rare and threatened species.

Jeff E. Herrick is a scientist with the U.S. Department of Agriculture–Agricultural Research Service Jornada Experimental Range, focusing on the factors controlling resilience at multiple scales and on the development of assessment and monitoring systems.

Richard L. Knight is professor of wildlife conservation at Colorado State University and coeditor of *The Essential Aldo Leopold* (1999) and *Aldo Leopold and the Ecological Conscience* (2002).

Rurik List is a conservation biologist at the Instituto de Ecología at the Universidad Nacional Autónoma de México, focusing on endangered species habitat conservation.

Mark Lockwood is the state natural-resource coordinator for the Trans-Pecos region in Texas. He leads river-rehabilitation efforts in Big Bend Ranch State Park.

Laura López-Hoffman is assistant professor at the School of Natural Resources and Environment and the Udall Center for Studies in Public Policy at the University of Arizona. Her research on the ecology and policy of managing transboundary systems emphasizes ecosystem services shared by the United States and Mexico.

Donna Lybecker is assistant professor of political science at Idaho State University, focusing on natural-resource conservation on the U.S.–Mexico border.

Carlos Manterola is president of Grupo Anima Efferus and conservation director of the Jaguar Conservancy. He has worked with the Commission for Environmental Cooperation Trilateral Committee on its shared species agenda and international protected areas.

Lourdes Martínez is a graduate student at the Instituto de Ecología of the Universidad Nacional Autónoma de México.

Emily D. McGovern is a researcher and editor at the Udall Center for Studies in Public Policy of the University of Arizona, focusing on social and environmental aspects of the U.S.–Mexico border region and on water policy.

Bonnie McKinney is wildlife coordinator for the CEMEX–Proyecto El Carmen Wildlife Area in Coahuila, Mexico.

Rodrigo A. Medellín has worked on the ecology and conservation of Mexican mammals for more than thirty years. He is senior professor of ecology at the Instituto de Ecología of the Universidad Nacional Autónoma de México.

Stephen P. Mumme is professor of political science at Colorado State University. He has consulted for the Commission for Environmental Cooperation on North American inland water management and is a long-time observer of environmental management along the U.S.–Mexico border.

Jesús Pacheco is an ecologist, a conservation biologist, and a graduate student at the Instituto de Ecología of the Universidad Nacional Autónoma de México.

Jane M. Packard is associate professor and director of the Biodiversity Stewardship Lab at Texas A&M University. She integrates cultural and biological perspectives on biodiversity conservation, emphasizing communication across disciplinary and physical boundaries.

Jessica Piekielek is a doctoral candidate in the Department of Anthropology at the University of Arizona. Her research in environmental anthropology focuses on the U.S.–Mexico borderlands.

Adrian Quijada-Mascareñas is adjunct professor at the School of Natural Resources and Environment, University of Arizona. He is a herpetologist with conservation experience on Mexican private land and *ejidos*, working with Indigenous communities in particular.

Richard Reading is founder and director of the Denver Zoological Foundation's Department of Conservation Biology, associate research professor at the University of Denver, and affiliated faculty at Colorado State University.

Brian P. Segee is a staff attorney with the Environmental Defense Center in Santa Barbara, California. He was previously staff attorney with Defenders of Wildlife in Washington, D.C., focusing on the protection of imperiled species and their habitat.

Christopher Sharp is a doctoral candidate at the University of Arizona, focusing on recreation behavior in backcountry and wilderness settings, measures of user impacts, and geographical information system techniques for recreation management.

Rodrigo Sierra Corona is a doctoral student at the Instituto de Ecología of the Universidad Nacional Autónoma de México.

Carlos Sifuentes is director of Áreas de Protección de Flora y Fauna Maderas del Carmen y Cañón de Santa Elena and leads river-conservation efforts along the Río Bravo in Mexico.

Joe Sirotnak is lead botanist at Big Bend National Park in Texas, responsible for vegetation management, exotic plant management, ecological restoration, and threatened and endangered plant species conservation.

Matt Skroch is a research assistant at the University of Arizona's School of Natural Resources and Environment studying conservation implications of climate change in the transboundary Madrean Archipelago.

Rachel Rose Starks is research analyst for the Native Nations Institute for Leadership, Management, and Policy at the University of Arizona's Udall Center for Studies in Public Policy. She has research experience in tribal governance, border issues, tribal courts, and asset building.

Robert G. Varady is deputy director of the University of Arizona's Udall Center for Studies in Public Policy, where he is research professor of environmental policy. Varady has written extensively on U.S.–Mexico environmental policy and global water initiatives.

Cora Varas is a doctoral candidate in the School of Natural Resources and Environment at the University of Arizona, researching genetic techniques to understand the evolution and habitat connectivity of southwestern black bear populations.

Evan R. Ward is associate professor of history at the University of North Alabama and author of *Border Oasis: Water and the Political Ecology of the Colorado River Delta, 1940–1975* (University of Arizona Press, 2003).

Francisco Zamora-Arroyo is director of the upper Gulf of California program at the Sonoran Institute and a researcher in the Department of Geosciences at the University of Arizona. He has worked in the Colorado River delta for more than ten years, building collaborations between water managers and local stakeholders.

Index

agave, 139, 142–43, 179–80, 184

Agreement Between the United States of America and the United Mexican States on Cooperation for the Protection and Improvement of the Environment in the Border Area. *See* La Paz Agreement

agriculture, 11, 12, 13, 14, 15, 28, 42, 60, 130n10, 143; Colorado River Delta and, 23, 25–26, 31–33; pest control, 180–82; prairie dogs and, 194–95

All-American Canal: water from, 25, 26, 28, 34, 139, 140–42, 145, 146–47, 148–49, 150

Andrade Mesa wetland, 25–26, 34, 140, 141, 148–49

Animal Damage Control Act, 87

Animas Foundation, 76

Apaches, 8, 11, 71, 72–73, 75

Arizona and California Tribal Liaisons, 65

Arizona Department of Environmental Quality, 63

Arizona Game and Fish Department, 107

Association of Fish and Wildlife Agencies, 183

Association of Zoos and Aquariums: Mexican wolf recovery, 119, 124

basketry, 61–62

bats, 145, 147, 173, 205, 261; conservation, 183–84, 267–68; ecosystem services, 170, 178–83; migratory, 81, 135–36, 137, 146(fig.), 172, 293; pollinating, 139, 142–43, 215, 228, 268; transboundary region, 171, 174–77(table)

Bavispe-Aros region, 93, 94

bears, black, 81, 85–87, 88–90, 246

BECC. *See* Border Environment Cooperation Commission

Bellagio Draft Treaty on Transboundary Groundwater, 148, 150

Big Bend National Park, 39, 44, 86

Big Bend Ranch State Park, 39, 44, 91

Big Bend reach (Rio Grande), 8, 39, 40–41, 43, 88, 91; protection, 44–45; restoration, 45–48, 294

Binational Chamber of Commerce, 252

binational institutions, 4, 8, 17–19, 26, 35, 138, 141, 142, 149–51, 165, 166, 248, 259, 261–66, 272–76, 281–84, 287–89, 292, 295, 297, 298–300. *See also* BECC; Bellagio Draft Treaty on Transboundary Groundwater; Border Industrialization Program; Border XXI; Border 2012; CEC; IBC; IBEP;